QUALITATIVE
INORGANIC ANALYSIS

QUALITATIVE
INORGANIC ANALYSIS

by

A. J. BERRY, M.A.

*Fellow of Downing College and University
Lecturer in Chemistry*

SECOND EDITION

CAMBRIDGE
AT THE UNIVERSITY PRESS
1948

CAMBRIDGE
UNIVERSITY PRESS

University Printing House, Cambridge CB2 8BS, United Kingdom

Cambridge University Press is part of the University of Cambridge.

It furthers the University's mission by disseminating knowledge in the pursuit of
education, learning and research at the highest international levels of excellence.

www.cambridge.org
Information on this title: www.cambridge.org/9781316509838

© Cambridge University Press 1948

First edition 1938
Second edition 1948
First paperback edition 2015

A catalogue record for this publication is available from the British Library

ISBN 978-1-316-50983-8 Paperback

CONTENTS

EXTRACT FROM THE PREFACE
TO THE FIRST EDITION

IT must be admitted that the teaching of analytical chemistry, particularly qualitative analysis, in this country has not progressed to anything like the extent to which the teaching of other branches of chemistry has advanced. Although much research work on analysis is published every year, the teaching of analytical chemistry is still largely carried on by old-fashioned methods. As regards qualitative analysis, there are even signs of a retrogression; the present-day student, although well versed in organic and in physical chemistry, knows less of qualitative analysis than his predecessor twenty-five years ago.

Former generations of students acquired a sound training in qualitative analysis with the aid of practical books such as Valentin's *Practical Chemistry* and Fenton's *Notes on Qualitative Analysis*, but such works are not well suited to present conditions of teaching. The modern student has so many demands made on him in the way of acquiring his chemical knowledge, that the attention which he can give to analysis is necessarily limited.

The present work has been written primarily in the interests of the author's own pupils, and in two directions radical departures from established custom have been made. In the first place, the ancient and arbitrary distinction between the common elements and a few of the so-called rare elements has been broken down. Anyone who cares to consult a chemical catalogue can see for himself that the prices of compounds of a number of elements, which are still described in many text-books as rare, are sufficient to condemn any such classification. In making a selection of such elements, their scientific interest and technical importance have been taken into consideration. In the second place, much progress has been made in the way of applying new reagents, particularly organic compounds, to inorganic analysis. Every year sees additions to a rapidly growing technique of qualitative analysis carried out by "drop reactions". Here, the selection of new tests was found to be much more difficult. The object in view has been to awaken the interest of the student in this aspect of the work by introducing him to a few of the more readily accessible new reagents, without neglecting the older methods which depend upon detailed separations.

PREFACE TO THE
SECOND EDITION

IN revising this work, the author has taken the opportunity of making various emendations, chiefly relating to the application of certain tests. In the first edition, the systematic analysis of the metals was discussed without using the traditional tabular methods of presentation because of the excessive reliance which many students are apt to place on such tables. It has, however, been recognized, as a consequence of representations from various friends, that this procedure has resulted in greater loss than gain to the average student, and accordingly analytical tables have been added towards the end of the book.

The author desires to express his sincere thanks for the valuable assistance which he has received from Dr F. Wild, Fellow and Tutor of Downing College, in connexion with the work of revision. He also desires to acknowledge with gratitude the ever helpful cooperation of the Staff of the University Press.

A. J. B.

CAMBRIDGE
Lent Term 1947

Part I
GENERAL

Chapter I

CERTAIN GENERAL PRINCIPLES

INTRODUCTION

THE objects of the two main divisions of chemical analysis, namely qualitative analysis and quantitative analysis, are sufficiently indicated by the terminology; the former being concerned with the detection of substances, the latter with their determination. In volumetric and in gravimetric analysis, the reactions which are employed for effecting the determinations must proceed completely. Such, however, is not the case in qualitative analysis. In many of the tests which are employed in qualitative work, it is of little or no consequence whether the reactions proceed on quantitative lines or not, or even if the products which are formed for the recognition of substances are of definite composition.

Although the objects and many of the methods involved in the two branches of chemical analysis are widely different, it frequently happens that one branch of analysis must supplement the other. Occasionally, it is not possible to identify a highly complicated compound by qualitative methods alone, and it will then become necessary to make quantitative determinations of its constituents. Much more frequently, qualitative tests are required to check the accuracy of quantitative experiments. In volumetric analysis, for example, the end-point in some titrations is found by colour changes which are really of the nature of microchemical tests; and in gravimetric analysis, recourse to qualitative tests is often required to ascertain the completeness of a separation or the purity of a substance which is weighed. Hillebrand and Lundell have urged that "a little less attention be paid to the method of weighing and a little more to the thing that is weighed".*

The General Functions of Acids and Bases

According to the elementary theory of ionization, acids are supposed to derive their distinctive properties from their capacity to yield free hydrogen ions, or more correctly, free hydroxonium, H_3^+O, ions. Similarly, the characteristic properties of bases are identified with their property of yielding free hydroxyl ions. A closer study of these two classes of compounds, however, has shown that the undissociated molecules of both acids and bases have a definite *role* to play, particularly when the effects of the presence of neutral salts are examined.

Acids and bases may be classified as strong or weak according to the extent to which their distinctive properties are exercised. In terms of the classical theory of ionization, strong acids and bases are highly ionized, whereas weak acids and bases are ionized only to a very limited extent. It might therefore appear at first sight that two acids of identical strength should be equally efficient agents in bringing a substance into solution. Such, however, is by no means necessarily the case. Hydrochloric and nitric acids are of practically identical strength, yet many metallic oxides are more readily dissolved by the former than the latter. In the case of ferric oxide, the more powerful solvent action of hydrochloric acid is to be attributed to the formation of complex ferrichloride ions.

When discussing the concentration of an acid or base in solution, two wholly different meanings may be attached to the term concentration. The term may refer to the total quantity of acid or base in a specified volume of the solution, without drawing any distinction between the undissociated molecules and the ions. Alternatively, reference may be restricted to the ionized

* *Applied Inorganic Analysis*, p. 3.

fraction, i.e. to the actual free hydrogen or hydroxyl ions. In order to avoid the use of negative indices, the concentration of hydrogen or hydroxyl ions in a solution is always expressed logarithmically in terms of the so-called pH scale. On this scale, a neutral solution (i.e. a solution in which the concentrations of hydrogen and hydroxyl ions are equal) has a pH value of 7,* acid solutions have pH values numerically less than 7, and alkaline solutions have pH values which numerically are greater than 7. In many analytical operations, it is important to keep the pH value of a solution within certain limits, as the following example will show.

The precipitation of metallic hydroxides is frequently effected by adding a weak base such as ammonia to the solution. If ammonia be added to a solution of a mixture of an aluminium and a magnesium salt, the hydroxides of both metals are precipitated. But if ammonium chloride followed by ammonia is added to the mixture, aluminium hydroxide alone will be precipitated. The addition of ammonium chloride limits the precipitating power of the base. This action, known as buffer action, is considered to be due to suppression of the hydroxyl ions from the ammonium hydroxide in consequence of the considerable increase in the concentration of the ammonium ions derived from the strong electrolyte ammonium chloride.

Examples of the buffering of weak acids by adding one of their salts are common in analytical work. Thus hydrogen sulphide will not precipitate zinc sulphide from a solution of zinc sulphate. But if sufficient sodium acetate is added to a solution of zinc sulphate, zinc sulphide may be precipitated completely by passing in hydrogen sulphide. The basic ferric acetate method of separating phosphoric acid depends on the same principle.

The acid or alkaline reaction of salts in solution arises as the result of hydrolysis, i.e. reaction between the ions of water with the ions of the salt, always tending to produce the least ionized product. Thus when sodium acetate is dissolved in water, the resulting solution has a pH value between 8 and 9, depending

* At room temperature.

on the concentration of the solution. The hydrogen ions derived
from the water combine with the acetate ions to produce the
feebly ionized acetic acid, the net effect of the equilibrium being
to leave a small but decided excess of hydroxyl ions in the
solution. Similarly, when ammonium chloride is dissolved in
water, the pH value of the solution usually lies between 4 and 5.
In this case, the hydroxyl ions derived from the water unite
with some of the ammonium ions to produce the feebly ionized
ammonium hydroxide, and equilibrium is established with a
small excess of hydrogen ions in solution. Solutions of salts of
which both cation and anion are constituents of strong bases
and acids are neutral, because hydrolysis does not arise to an
appreciable extent.

It may be added that some of the fundamental conceptions
of acids and bases have undergone modifications in recent years.
It is held by some chemists that the distinctive properties of
weak acids and weak bases are to be considered as due not to
the limited ionization of these substances, but to equilibria
between pseudo acids and true acids and between pseudo bases
and true bases, the true acids and bases being completely
ionized. The apparent weakness is considered to be due, not to
the intrinsic properties of the compounds, but to the extremely
small concentration of hydrogen or hydroxyl ions which is in
equilibrium with the pseudo form. In this connexion, it has
been pointed out by Thiel that it is incorrect to describe carbonic
acid as an extremely weak acid. The error has arisen as a consequence of regarding dissolved carbon dioxide as consisting
wholly of carbonic acid and not partly of carbonic anhydride.
Equilibrium between carbonic acid and its ions is established
immediately, but the hydration of dissolved carbon dioxide is a
time reaction. When the true ionization constant of carbonic
acid is measured, the acid is estimated to be about twice as
strong as formic acid.

An extended conception of the fundamental ideas about acids
and bases was introduced by Brønsted and by Lowry in 1923,
who defined acids as substances which can lose hydrogen ions

and bases as substances which can acquire hydrogen ions. In this definition, Brønsted refers to the unhydrated hydrogen ion, the proton. It will be evident that this way of regarding acids and bases extends these terms to include substances which are not defined as acids or bases in the ordinary way, although their acidic or basic properties are none the less recognizable.*

The Potential or Electrochemical Series of the Metals

It is well known that metals differ very widely as regards their reactivity with acids. Thus with hydrochloric acid, magnesium readily dissolves with evolution of hydrogen in the dilute acid, zinc dissolves more slowly, indeed very slowly if the metal is fairly pure, while tin is practically without action upon the dilute acid, but dissolves in the concentrated acid on heating. Relevant to this behaviour is also the displacement of one metal from a solution of its salts by another metal. Thus zinc displaces copper from a solution of copper sulphate, and copper displaces mercury from a solution of mercurous nitrate. Such phenomena are due primarily to differences in the tendency of the metals including hydrogen, which in this respect resembles metals, to become ionized. When a metal is placed in a solution of its own ions, the tendency to assume the ionic condition is a measurable quantity which can be expressed in volts. This tendency is greatest with the alkali metals, and least with gold and platinum. If the metals are arranged in the order of ascending normal potential, as determined by electrochemical measurements, it will be found that the same order is followed with regard to diminution of electropositive character. The commoner metals become arranged in the following order:

K, Ca, Al, Mn, Zn, Fe, Cd, Co, Ni, Sn, Pb, H, Cu, Hg, Ag, Au.

A knowledge of the relative order of the metals in the potential series is of value in studying their analytical reactions, but it must be emphasized that the chemical behaviour of any parti-

* See *Inorganic Chemistry* by Niels Bjerrum, English edition, 1936, pp. 107 *et seq.*

cular metal may be much more complicated than its relative position in the potential series might at first sight appear to indicate. In this connexion the behaviour of certain metals, particularly zinc with dilute hydrochloric or sulphuric acid, is of much interest. Pure zinc is remarkably resistant to the attack of these acids, which may at first sight appear surprising, as the metal stands considerably above hydrogen in the potential series. The impure metal dissolves rapidly, and the pure metal also can be made to dissolve with equal rapidity if it is touched with a piece of silver or platinum under the surface of the acid. A completed Voltaic circuit is necessary for the metal to dissolve in and displace hydrogen from the acid. The action of copper upon acids is also instructive. As this metal follows hydrogen in the potential series, it would not be expected to displace hydrogen from acids, but only to be attacked by oxidizing acids such as nitric acid or by hot concentrated sulphuric acid, in which it dissolves with formation of reduction products of these acids. Copper is, however, dissolved when heated with hydriodic or hydrobromic acid with evolution of hydrogen, and if the experiments are carried out quantitatively, it will be found that the volume of hydrogen which is evolved corresponds with the *cuprous* equivalent of the metal. The action of these acids in dissolving the metal is due to the strong tendency of copper to form complex halogen anions of the type $\overline{CuX}_n$. If the solution obtained by the action of hydrobromic or hydriodic acid on copper is diluted with water, the complex anion is decomposed with separation of cuprous bromide or iodide. Even boiling concentrated hydrochloric acid has an appreciable solvent action upon copper, as may be seen by applying sensitive tests. The action of an acid upon a metal is determined not only by the tendency to discharge hydrogen ions, but to the capacity for producing complex ions.

OXIDATION AND REDUCTION

The fundamental conception of oxidation is the addition of oxygen to a substance or the removal of hydrogen from it. In a similar sense, the term reduction expresses the addition of hydrogen to a substance or the removal of oxygen from that substance. The meaning of the terms has been extended beyond the original conceptions by considering other electro-negative elements as functioning similarly to oxygen, and electropositive elements as behaving similarly to hydrogen. Electrochemical definitions of the terms have been formulated in this more extended sense. Thus oxidation is frequently defined as increasing the ratio of the electronegative part to the electropositive part of a molecule, or diminishing the ratio of the electropositive to the electronegative part of the molecule. Reduction is similarly defined in reciprocal terms.

Remy has pointed out that oxidation and reduction when considered in the purely chemical sense may be in conflict with electrochemical definitions. Thus when calcium hydride is dissociated by heat, hydrogen is expelled. Is this change to be regarded as an oxidation or as a reduction? From the purely chemical standpoint, hydrogen is removed from the substance and therefore the change must be regarded as an oxidation. But since experiments on the electrolysis of calcium hydride have shown conclusively that hydrogen is the electronegative constituent of this compound, the chemical change of thermal dissociation must be regarded as a reduction when the electrochemical definitions of oxidation and reduction are taken as the standards. Rigid definitions of fundamental conceptions in chemistry are very difficult to formulate without complications arising as the science develops.

Reactions which involve oxidation and reduction in aqueous solution are sometimes accompanied with characteristic changes of colour. In subjecting a substance to a preliminary examination in qualitative analysis, it is frequently desirable to investigate its response to the action of oxidizing and reducing agents, and to note any colour changes. For example, zinc and dilute sulphuric acid will convert dichromates (orange) into green chromic and ultimately into the blue chromous salts. Under similar conditions, titanic salts which are colourless are converted into the violet titanous salts. Derivatives of quinquevalent vanadium (vanadic acid) undergo an interesting series of colour changes as reduction proceeds from a yellow colour through blue (vanadyl salts), then green (vanadic salts) and ultimately to the violet colour of vanadous salts.

Reactions which are concerned with oxidation and reduction can, of course, be expressed by ordinary chemical equations. An older way of doing this is to give expression merely to changes in the states of oxidation. Thus the oxidation of iron from the ferrous to the ferric condition by potassium dichromate in acid solution may be expressed by the "oxide" equation:

$$6FeO + 2CrO_3 = 3Fe_2O_3 + Cr_2O_3,$$

or by an equation which is concerned with the resulting changes of valency:

$$3Fe(II) + Cr(VI) = Cr(III) + 3Fe(III).$$

Both the "oxide" and the "valency" equations are accurate in so far as they furnish a quantitative statement of the actual state of affairs. They are however incomplete in one important particular. Free hydrochloric or sulphuric acid is concerned in the reaction, or, more generally, hydrogen ions are involved. The reaction may therefore be written:

$$\overset{--}{Cr_2O_7} + 6\overset{++}{Fe} + 14\overset{+}{H} = 7H_2O + 2\overset{+++}{Cr} + 6\overset{+++}{Fe}.$$

Substances may be classified as oxidizing or as reducing agents according to their behaviour with certain types of substances. Some compounds can function as oxidizing agents in some reactions and as reducing agents in others. Thus nitrous acid would be described as an oxidizing agent when it reacts with potassium iodide, since iodine is liberated, but nitrous acid will decolorize potassium permanganate, and in those circumstances it would be classed as a reducing agent. Hydroxylamine normally

CATALYSED AND INDUCED REACTIONS 9

acts as a powerful reducing agent, but it can also oxidize freshly precipitated ferrous hydroxide to ferric hydroxide. In acid solution the reaction is reversed, ferric salts being reduced to the ferrous condition.

Catalysed and Induced Reactions. Numerous reactions which involve oxidation and reduction are profoundly influenced by the presence of traces of substances which exert a catalytic action. A test for tartrates depending on the catalytic action of a trace of ferrous ion was discovered many years ago by Fenton. In the presence of this catalyst, tartaric acid is oxidized to dihydroxymaleic acid by hydrogen peroxide, whereas, without a trace of ferrous salt, no obvious oxidation takes place. The course of the oxidation of thiosulphates by hydrogen peroxide has been shown by Abel to be capable of variation according to the experimental conditions. In dilute acetic acid solution, particularly in presence of a trace of iodide ion, oxidation proceeds to the tetrathionate stage, whereas in neutral solution with a trace of ammonium molybdate as catalyst, the oxidation proceeds to sulphate. More recently, Feigl has devised a useful test for silver which depends upon the catalytic action of the metal on the reduction of a ceric salt (see p. 30).

Induced reactions resemble catalytic reactions in many respects. A solution of sodium sulphite is slowly oxidized to sulphate by exposure to atmospheric oxygen. A solution of sodium arsenite, on the other hand, is remarkably stable to atmospheric oxygen. But if air is passed through a solution of the two salts, *both* are oxidized. Many similar examples are known. A most interesting application of this principle has been applied by Feigl and Krumholz to increasing the sensitiveness of the alkaline stannite test for bismuth. Bismuth salts are rapidly reduced to the metallic condition by a solution of sodium stannite (see p. 40). The sensitiveness of the test may, however, be increased one hundredfold as follows. Lead salts are reduced to the metallic state very slowly by sodium stannite. In the presence of a concentration of one part of bismuth in five million, the lead salt is rapidly reduced.

COLLOIDAL PHENOMENA IN ANALYSIS

When substances react together in solution in such a way as to form an insoluble product, it sometimes happens that a precipitate is not produced at once. This may arise as a result of supersaturation, or it may be due to the substance being held in colloidal solution. Examples are frequently to be encountered in analytical work. Thus when hydrogen sulphide is passed through a solution of arsenious acid, the solution becomes yellow, but arsenious sulphide is not actually precipitated unless hydrochloric acid or some other suitable electrolyte is present. The yellow liquid which is produced in the absence of electrolytes contains arsenious sulphide in colloidal solution. Under some conditions aluminium hydroxide may remain partly in colloidal solution when ammonia is added to a solution of an aluminium salt in the absence of ammonium salts.

When some insoluble precipitates are washed for a long time with water, a point may be reached when there is a marked tendency for the insoluble substance to pass through the filter. If the washing is conducted with a dilute solution of an electrolyte instead of with water, this does not arise. If thallous iodide is washed with water, after some time the liquid assumes a pale yellow colour. This phenomenon is known as peptization, and is due to the solid passing into colloidal solution. If the washing is conducted with a dilute solution of potassium iodide, the washings remain colourless.

Colloids are usually classified broadly into two types, formerly termed suspensoids and emulsoids, and now usually distinguished as lyophobic colloids and lyophilic colloids. In the former type, there is no tendency for the dispersed substance to unite with the "solvent", whereas in the latter type there is a more or less strong tendency for the colloidal substance to enter into a somewhat ill-defined combination with the solvent. The distinction between lyophobic and lyophilic colloids is, however, not a rigid one.

Colloidal solutions differ from true solutions in various respects. In consequence of the very small concentration of the dissolved substance the osmotic pressure of the solution is extremely small, and the same is true of the other properties which depend upon osmotic effects. Although a colloidal solution may appear homogeneous to the eye, examination by ultra-microscopic methods shows such solutions to be really heterogeneous, since the ultra-microscopic particles exert a scattering action upon light. The most characteristic feature of colloidal solutions as distinct from true solutions is the non-diffusibility of colloids. Many fundamental experiments were carried out by Graham, who developed the method of separating colloids from crystalloids by dialysis, i.e. effecting separation by causing the substances in true solution to pass through a suitable membrane and at the same time retaining the colloid within it.

The particles in a colloidal solution are electrically charged, and the presence of the charge exerts a stabilizing effect on the solution. If the charge is removed, the colloid becomes unstable and flocculation or gelatinization takes place. The sign of the charge on the particles can be determined by placing two electrodes maintained at a fairly high difference of potential in the colloidal solution, and observing whether the particles wander towards the anode or the cathode. The sign of the charge on the particles is determined, not only, as is sometimes supposed, on the nature of the substance, but also on the way in which the colloidal solution has been prepared. Fundamental experiments on the formation of hydrosols were carried out by Lottermoser in 1905, who showed that colloidal solutions of a substance such as silver iodide could be prepared with the colloidal particles charged positively if the solution was prepared with a slight excess of silver ions, or negatively if the halogen ions were in excess.

The flocculating action of electrolytes on colloidal solutions is a function of the valency of the ions. Tervalent ions are much more efficient than bivalent ions, and these are considerably more efficient than univalent ions. If two colloidal solutions in which the particles are charged with the same sign are mixed together, flocculation will not take place, but if oppositely charged colloidal solutions are mixed, flocculation takes place at once with separation of both substances. Thus arsenious sulphide and antimonious sulphide as usually prepared are negative colloids, and ferric hydroxide as usually prepared is a positive colloid. The colloidal solutions of antimonious sulphide and arsenious sulphide may be mixed together without precipitation occurring, but if the colloidal solution of ferric hydroxide be added to either solution, precipitation takes place.

Colloids have considerable capacity for adsorbing substances from solution. The electric charges associated with substances in colloidal solution are most probably due to preferential adsorption of ions. Many so-called adsorption compounds are highly coloured. Thus the intensely blue substance which iodine forms with starch is most probably an adsorption product. Bunsen discovered that freshly precipitated ferric hydroxide was an efficient antidote for poisoning by arsenious acid. He considered that insoluble ferric arsenite was produced. Subsequent experiments have, however, shown that the ferric hydroxide hydrogel removes the arsenious acid from solution by adsorption.

Chapter II

THE METHODS OF QUALITATIVE ANALYSIS

THE SENSITIVENESS OF QUALITATIVE TESTS

IN ordinary work in qualitative analysis, precipitation reactions and colour tests are most frequently employed for identifying substances. Occasionally other properties are utilized, e.g. catalytic action. If a substance is known to be a specific catalyst for some particular reaction, it is possible to carry out the reaction with and without the substance which has to be tested, and observe if the reaction is accelerated. It is much easier to express the sensitiveness of colour tests, and, it may be added, of catalytic tests, in numerical terms than the sensitiveness of precipitation reactions, because the latter are much more subject to interference by small changes in the experimental conditions.

Quantitative expression of sensitiveness may be given in terms of actual mass of substance detectable or as a function of concentration. The well-known nitroprusside reaction for sulphides is stated to be sensitive to 10^{-6} g. of sodium sulphide, or, alternatively, to be perceptible with a concentration of one part of sodium sulphide in 50,000. It is instructive to compare the sensitiveness of a few colour reactions with other means of detecting traces of substances. According to Bunsen and Kirchhoff, the smallest quantity of sodium which can be detected with the aid of the spectroscope is 3×10^{-10} g. Modern spectroscopic methods have exceeded this degree of sensitiveness, but when compared with electroscopic methods of detecting radioactive substances, a striking contrast can be observed. Quantities of material of the order of 10^{-15} to 10^{-17} g. have been detected by Paneth using electroscopic methods; in particular the amount of the hydride of the radioactive isotope of bismuth which was first detected was actually about 10^{-15} g. Very small quantities of certain substances can be detected by their characteristic odours. As the result of an investigation carried out in 1887, Emil Fischer and Penzoldt concluded that the odour of mercaptan was some 250 times more delicate as a test for this compound than was the spectroscopic test for sodium as determined by Bunsen and Kirchhoff. It has been estimated that quantities of silver of the order of 10^{-8} g. are sufficient to inhibit the growth of certain micro-organisms. The limits of so-called purely chemical methods of detection would seem to be of the order of 10^{-8} g. Thus it is possible to detect this quantity of silver by observing its catalytic reduction on ceric salts (see p. 30). A similar quantity of nitrous acid can be detected by a reaction which depends upon diazotization followed by coupling (see p. 91), and the limiting concentration is one part of nitrous acid in five million.

General and Specific Reagents

In analytical chemistry, a distinction is drawn between what are termed *general* reagents and *specific* reagents. The distinction is perfectly simple and obvious. Hydrogen sulphide would be termed a general reagent, because it can effect the separation of a considerable number of metals as insoluble sulphides. Ammoniacal silver nitrate would similarly be termed a general reagent for reducing substances, because many readily oxidizable substances react with it with separation of silver. On the other hand, potassium ferricyanide would rightly be classed as a specific reagent for ferrous salts, since no other substances react with this compound to produce the well-known blue precipitate of ferrous ferricyanide (Turnbull's blue). Similarly, potassium xanthate would be characterized as a specific reagent for molybdates (see p. 49).

It is frequently possible to adopt devices with general reagents with the object of limiting their reactivity, and thereby making their action more specific. Silver nitrate has been described as a group reagent for acid radicals, because it is capable of precipitating a considerable number of acids as sparingly soluble silver salts. But if silver nitrate is applied in presence of nitric acid, the only insoluble precipitates which are produced are those derived from chlorides, bromides, iodides, and certain cyanogen acids. One method of limitation which is frequently adopted is that of complex ion formation. The detection of cadmium in the presence of copper is usually effected by adding excess of potassium cyanide to the solution followed by hydrogen sulphide. No cupric sulphide is precipitated because the copper is present as an extremely stable anion, whereas the cadmium complex anion is unstable, and yields a sufficient concentration of cadmium ions for precipitation to take place.

Great care should be observed in describing any particular reagent as specific for some one element or compound unless the most careful scrutiny has been made of its behaviour under varied experimental conditions. Uncertainties in practical work may be minimized by making blank tests and by using alternative reagents. Much progress has been made in recent years in the way of applying new reagents, particularly certain organic compounds, in the qualitative analysis of inorganic substances. Many of these compounds are either definitely specific reagents for particular substances, or their mode of action can be varied in some way with the object of effecting limitations. In the older methods of qualitative analysis of a mixture of substances, separations of the metals, first by the use of so-called group reagents, and afterwards by separations within the groups, was the universal practice. At the present time, methods have been described for identifying the constituents of fairly complicated mixtures which involve no separations but depend solely on the use of these newer reagents. As many of these reagents are extremely sensitive, it is possible to effect a complete analysis with very small quantities of material, the solutions being tested with single drops. Whether or not it is advantageous to carry out any particular analysis in the older way by making separations or in the modern way by "spot tests" is a question to which no general answer can be given at present: much depends upon the nature of the substance which is being analysed. In most cases, however, it will almost certainly be advisable to effect separations by group reagents in the first instance before proceeding to investigate the effects of applying "spot tests" for particular substances.

Recently, 1937, a Committee appointed by the International Union of Chemistry to make a critical study of new analytical reagents decided to differentiate between *specific* and *selective* reagents. It is recommended that the term *specific* should be reserved for such reagents and reactions which under the particular experimental conditions employed are capable of indicating one substance only, whereas those which are characteristic of a relatively small number of substances should be described as *selective*.

Much progress has been made in microchemical analysis in consequence of the introduction of special apparatus for handling very small quantities of substances. Descriptions of the apparatus are to be found in larger works, such as Emich's *Lehrbuch der Mikrochemie* and Feigl's *Qualitative Analyse mit Hilfe von Tüpfelreaktionen*. For ordinary work involving spot tests, however, there is no particular necessity to be provided with any apparatus beyond a few glass rods, small pipettes or dropping tubes, a porcelain plate or white tile, a few small porcelain crucibles, and a plentiful supply of good filter paper.

APPLICATION OF THE IONIC THEORY TO ANALYTICAL PROCESSES

The separation of the metals into some five or six groups by applying successively reagents which effect separation by the production of insoluble precipitates had been practised long before a general theory of analytical processes was given by Ostwald some forty years ago. The outlines of this theory, based upon the classical theory of electrolytic dissociation, may now be briefly explained. When an electrolyte is dissolved in water, there is an equilibrium between the undissociated molecules and the ions, the degree of ionization depending upon the concentration of the solution and the temperature. Strong acids and bases, and the great majority of salts, are highly ionized at moderate concentrations, whereas weak acids and bases are only very slightly ionized. Mercuric salts are ionized to an extremely limited extent, and the same is true, though to a smaller degree, for a few other salts, such as cadmium salts and thallic salts.

If for simplicity we consider the case of a binary electrolyte which dissociates into two univalent ions, let a denote the concentration of the cation, b that of the anion, and c that of the undissociated molecules; it follows from the law of mass action that $ab = kc$, where k is a constant which depends upon the particular substance and on the temperature. When the solution is saturated, the value of the product ab is known as the solubility product, and if this value is exceeded, precipitation will take place. Increase in the value of the product ab and consequently of c may be attained by increasing the concentration of either of the ions. The precipitation of sparingly soluble substances from solution by adding a solution of an electrolyte having an ion in common is therefore explained in a simple manner.

Just as the production of a sparingly soluble precipitate is explained by the over-stepping of the solubility product, so the phenomenon of the dissolution of a precipitate by an acid receives a similar explanation. Salts of weak acids, such as calcium oxalate, dissolve in strong acids because of the tendency to form the less ionized weak acid. The addition of hydrogen ions has the effect of withdrawing the anions of the weak acid, with the result that the previously saturated solution of the nearly insoluble salt now becomes unsaturated, because the solubility product is no longer reached. This process will continue until the precipitate is dissolved.

It frequently happens that when a reagent is added to a solution of an electrolyte, precipitation takes place, but further addition of the reagent results in the precipitate being redissolved. Thus when ammonia is added to a solution of cupric sulphate, a precipitate of cupric hydroxide is first produced, but on further addition of the reagent, a deep blue solution is obtained. Phenomena of this kind are always due to the formation of complex anions or cations; in the particular example just mentioned, the solution contains the copper as a complex cation, having the formula $Cu(\overset{++}{N}H_3)_4$.

The application of the classical theory of electrolytic dissociation to analytical processes has doubtless resulted in a fairly clear understanding of many of the operations which are regularly carried out in this kind of work, but it must be emphasized that certain difficulties still remain. For example, the sulphides of nickel and cobalt are not precipitated from solutions of these salts when hydrogen sulphide is passed through them, although they are readily precipitated from alkaline solutions. On the other hand, when obtained by precipitation with ammonium sulphide, nickel and cobalt sulphides are not dissolved by dilute hydrochloric acid. It would appear that insoluble precipitates are frequently highly polymerized molecules, the properties of which have become considerably altered as the result of polymerization. Freshly precipitated aluminium hydroxide is very easily dissolved in hydrochloric acid, whereas if the precipitate has been dried it is much less easily dissolved, and ignited aluminium oxide is extremely resistant to the attack of acids.

In recent years, the classical theory of electrolytic dissociation has gradually been abandoned as far as the behaviour of strong electrolytes is concerned. Strong electrolytes are now regarded as being completely dissociated at all concentrations, and the examination of crystalline substances by X-ray analysis has led to the conclusion that aggregations of ions are to be found in crystals. Some definite modification of the theory of analytical reactions to harmonize with these more modern views is therefore to be expected.

It will be seen in what follows how the principles which have been briefly discussed are applied practically to the division of the metals into groups in qualitative analysis.

19

PREPARATION OF A SOLUTION FOR ANALYSIS

A solution of the substance is prepared for the analysis. If the substance is soluble in water, an aqueous solution is prepared and examined directly. If it is insoluble in water, it must be brought into solution by other means. Try on separate small portions of the substance in the following order, *dilute hydrochloric acid, concentrated hydrochloric acid, nitric acid,* and *aqua regia.* When aqua regia is used, the ratio of hydrochloric acid to nitric acid should be at least four parts of the former to one of the latter.

If the substance fails to dissolve in any of the above-named reagents, it is usually necessary to bring it into a soluble condition by fusing it with a suitable flux. Thus insoluble silicates are opened up by fusing an intimate mixture of the silicate with about six times its weight of *anhydrous sodium carbonate* in a platinum crucible (see p. 101). Ignited metallic oxides, such as alumina or ferric oxide, and many minerals may be attacked by fusing them with *potassium bisulphate* in a silica crucible. Some chemists employ *sodium carbonate* mixed with an oxidizing agent, such as *potassium nitrate* or *sodium peroxide,* as a flux for the fusion of refractory minerals. Nickel crucibles are useful for this kind of work, as apart from being much cheaper than platinum vessels, they are less attacked by alkalis. Some nickel, however, is always brought into the melt.

Alloys are usually best dissolved in nitric acid, diluted with one to two parts of water. All the common metals, except tin and antimony, are thereby converted into nitrates; tin and antimony being oxidized into sparingly soluble hydrated oxides. Occasionally other reagents may, with advantage, be employed. Thus aluminium alloys which contain only small quantities of other constituents may be dissolved in aqueous sodium hydroxide, and the insoluble residue examined separately.

If the presence of organic matter has been indicated in a preliminary examination carried out by dry tests, it is desirable to destroy it by ignition, and to dissolve the ignited residue in

2-2

hydrochloric acid for the analysis. Some organic acids, such as tartrates, interfere with the precipitation of certain metals by ammonia, owing to the formation of complex ions. Oxalates also may cause complications. It may be added that a preliminary examination by dry methods should not be omitted.

As regards the quantity of material which should be taken for an analysis no general rules can be given. If the object in view is merely to identify a single substance, a few cubic centimetres of a solution of the order of two per cent concentration will usually be found suitable, but much more dilute solutions are desirable in special cases. If, however, the object in view is to test for traces of impurities in some industrial product, it is usually desirable to take a relatively larger quantity of the material, so that small quantities of impurities may be tested for without difficulty. In general, however, it is desirable to conduct an analysis with as small a quantity of material as may be convenient for the experiments which are to be performed.

Having obtained a solution of the substance, excess of *dilute hydrochloric acid* is added to it. This results in the precipitation of the metals of the *first* group. The metals of this group, viz. *silver, mercury (ous)*, and *lead* are precipitated as *chlorides*. Other metals which may appear in this group are *thallium (ous)* as *chloride* and *tungsten* as *tungstic acid*. The precipitated metals in this and in the subsequent groups are separated as described in Chapter V and identified by applying confirmatory tests.

The filtrate from the first group is heated and slowly saturated with *hydrogen sulphide* for about ten minutes. This results in the precipitation of the elements of the *second* group, viz. *lead, mercury (ic), bismuth, copper,* and *cadmium*. These metals constitute the *first* division of the group, the sulphides of which are insoluble in *yellow ammonium sulphide*. In addition to the above, the sulphides of the *second* division of the group, viz. the sulphides of *arsenic, antimony,* and *tin* are also precipitated. *Molybdenum sulphide* also belongs to this division of the group. These compounds all dissolve in yellow ammonium sulphide forming complex salts.

Having ascertained that the metals of the *second* group have been precipitated *completely* by hydrogen sulphide, the filtrate is boiled until hydrogen sulphide has been *completely* expelled, a few drops of nitric acid being added and the boiling continued for a short time. The object of this is to oxidize any ferrous iron to the ferric condition. Ammonium chloride followed by a *slight* excess of ammonia are added to the solution. The metals of the *third* group are then precipitated as hydrated oxides, viz. *aluminium, chromium*, and *iron*. Other elements which may be precipitated as hydrated oxides are *beryllium, titanium, cerium, thorium*, and *zirconium*. *Uranium* if present is precipitated as ammonium diuranate. The presence of certain anions, particularly phosphates, may cause serious complications in this group (see p. 123); insoluble phosphates being precipitated together with the metallic hydroxides. Indeed, the separations in the third group are by no means satisfactory; as metals belonging to subsequent groups are liable to be precipitated.

The filtrate from the third group is now treated with ammonium sulphide. This results in the precipitation of the metals which constitute the *fourth* group, viz. *cobalt, nickel, manganese*, and *zinc*, as sulphides. *Vanadium* may be present in the solution, but is not precipitated. It may, however, be precipitated as the sulphide by acidifying the filtrate.

The filtrate from the fourth group, which contains ammonia and ammonium salts, is treated with ammonium carbonate and warmed. *Barium, strontium*, and *calcium* are precipitated as carbonates. These elements constitute the *fifth* group.

The remaining elements constitute the *sixth* group. Of these, *magnesium* is precipitated as the ammonio-phosphate by adding sodium phosphate. The *alkali metals* are identified by special tests.

Notes on Certain Special Reagents

A considerable number of new reagents have found their way into inorganic qualitative analysis in recent years. A certain number of these are specific for certain metals, but, as has been mentioned already, too much reliance must not be placed upon their specificity until their behaviour has received an exhaustive study. It has been established beyond doubt that certain groupings exhibit a specific response to certain metals: for example, the dioxime group, $\begin{array}{c}-\text{C}=\text{NOH}\\ |\\ -\text{C}=\text{NOH}\end{array}$, has been stated to be specific for nickel, the reagent which is commonly employed being the compound dimethylglyoxime, $\begin{array}{c}\text{CH}_3-\text{C}=\text{NOH}\\ |\\ \text{CH}_3-\text{C}=\text{NOH}\end{array}$. Even in this well-established case, however, it is scarcely accurate to describe the reagent as specific for nickel, because although nickel forms a very sparingly soluble scarlet derivative, an intensely red soluble ferrous compound, having a similar constitution, is readily produced; indeed, it is a very useful test for small quantities of ferrous salts. The properties and application of a few of the more modern reagents may now be discussed. For a more exhaustive discussion, we may refer the reader to Feigl's *Qualitative Analyse mit Hilfe von Tüpfelreaktionen.*

1. *Ammonium mercurithiocyanate*, $(NH_4)_2Hg(SCN)_4$. This reagent is prepared by dissolving 8 g. of mercuric chloride and 9 g. of ammonium thiocyanate in 100 cc. of water. Its characteristic properties are those of the complex anion $[\overline{Hg(SCN)_4}]$. Certain metals produce sparingly soluble mercurithiocyanates, some of which possess distinctive colours. Thus cobalt salts produce a fine blue crystalline precipitate having the formula $Co[Hg(SCN)_4]$. Cadmium and zinc salts produce white precipitates, having similar formulae.

2. *Dimethylglyoxime* or *diacetyl-dioxime*, $\begin{array}{c}\text{CH}_3-\text{C}=\text{NOH}\\ |\\ \text{CH}_3-\text{C}=\text{NOH}\end{array}$, was introduced by Tschugaeff in 1905 as a reagent for nickel, with which it forms a very sparingly soluble scarlet derivative having the formula $\begin{array}{ccc}\text{CH}_3.\text{C:NOH} & & \text{ON:C.CH}_3\\ | & \diagdown\diagup & |\\ & \text{Ni} & \\ | & \diagup\diagup\diagdown & |\\ \text{CH}_3\text{C:NO} & & \text{HON:C.CH}_3\end{array}$. The compound is used for determining the metal gravimetrically. The reagent is employed in alcoholic solution of one per cent concentration, and ammonia must be added to the solution to be tested to neutralize the acid which is formed by the replacement of the hydrogen

atoms from the reactive oxime groups by the metal. Dimethyl-glyoxime is also used for the gravimetric determination of palladium. The colour reaction with ferrous salts mentioned above has been adapted as a test for tin. The substance to be tested is treated with a *ferric* salt and dimethyl glyoxime. Ferric salts do not react with this compound, but the ferrous salt which is produced by reduction gives rise to an intense red colour. It is clear that this is not a specific test for tin, as the reaction would be given by any substance which exerts a reducing action upon a ferric salt.

3. *Diphenylcarbazide,* $CO\begin{smallmatrix}\diagup NH.NH.C_6H_5\\ \diagdown NH.NH.C_6H_5\end{smallmatrix}$, is employed in a one per cent alcoholic solution as a reagent for cadmium and mercuric salts. Coloured derivatives of uncertain constitution are produced with these metals, that obtained with cadmium having a reddish violet colour, while the mercury compound is bluish violet. Certain anions having oxidizing properties, particularly chromates, also produce coloured products with this substance. The solution must be freshly prepared, as it rapidly develops a red colour on keeping. Exceptional care should be taken when making blank tests with this reagent.

4. *α-Nitroso-β-naphthol,* ⬡⬡OH, has long been known as
   NO
a sensitive reagent for cobalt, with which it forms a so-called

inner complex compound having the formula ⬡⬡O$\frac{Co}{3}$. The
   NO
reagent is usually prepared by dissolving 1 g. of α-nitroso-β-naphthol in 50 cc. of glacial acetic acid and diluting the solution with water to 100 cc. The solution to be tested should be nearly neutral or slightly acid. Coloured derivatives are formed with α-nitroso-β-naphthol and ferric, uranyl, and cupric salts, hence the reagent must not be considered as specific for cobalt.

5. *Salicylaldoxime,* $C_6H_4\begin{smallmatrix}\diagup CH:NOH\\ \diagdown OH\end{smallmatrix}$, has been introduced for

detecting copper and also for determining this metal gravimetrically. The compound which is produced is a pale greenish precipitate, having the formula $\left(C_6H_4 \underset{O-}{\overset{CH:NOH}{\diagdown}} \right)_2$ Cu, and it is insoluble in dilute acetic acid. With the exception of palladium, other metals form derivatives with this reagent which are soluble in dilute acetic acid. It is therefore scarcely accurate to speak of this reagent as specific for copper. The reagent is prepared by dissolving 1 g. of salicylaldoxime in 5 cc. of alcohol, and pouring the solution very slowly into 95 cc. of slightly warm water. The oily suspension which is thus produced gradually becomes clear when the liquid is shaken.

6. *Paranitrobenzeneazoresorcinol*, $NO_2.C_6H_4.N_2.C_6H_3(OH)_2$, is employed in very dilute alkaline solution (1 milligramme of the compound in 100 cc. of $2N$ sodium hydroxide) as a reagent for magnesium. The solution in alkali is a violet colour, but in presence of a trace of magnesium a blue colour or precipitate is produced. The blue colour is probably an adsorption product. The reaction is not specific for magnesium as certain other metals are capable of giving rise to coloured adsorption products. Ammonium salts must be removed before applying the test.

7. *Potassium xanthate*, $SC \underset{OC_2H_5}{\overset{SK}{\diagup}}$, is a specific reagent for molybdates, with which it forms an intensely coloured red purple compound when dilute hydrochloric acid is added. The derivative which is formed is said to possess the formula $MoO_3[SC(SH)(OC_2H_5)]_2$. The reagent is added in the solid condition to a drop of the solution to be tested and two drops of $2N$ hydrochloric acid are added.

8. *Dihydroxytartaric acid*, $\underset{C(OH)_2COOH}{\overset{C(OH)_2COOH}{|}}$, is a valuable reagent for the detection of sodium, with which it forms a very sparingly soluble sodium salt. Precipitation is more complete if the reaction is carried out in the presence of potassium carbonate to neutralize the resulting acid. Lithium salts produce a similar precipitate, but lithium dihydroxytartrate is more soluble than

the sodium salt. A volumetric method for determining sodium, based on the quantitative oxidation of the acid by potassium permanganate in presence of dilute sulphuric acid, has been devised.

9. *Phosphomolybdic acid*, H_3PO_4, $12MoO_3$, a compound in which the molybdenum is contained in the co-ordinated condition in the molecule, is a much stronger oxidizing agent than a simple molybdate. It is useful in testing for certain types of reducing agents in "spot" reactions, a deep blue reduction product, so-called molybdenum blue, being produced. The ammonium salt may sometimes be used, and is easily prepared by the well-known ammonium molybdate reaction for a phosphate, but it must be thoroughly purified from nitric acid before use. The free acid may be prepared from the ammonium salt by dissolving it in boiling aqua regia, evaporating to dryness, and recrystallizing the residue twice from water.

10. *Alizarin and other derivatives of anthraquinone.* Highly coloured derivatives of certain metals are produced with alizarin and with certain of its substitution products. Some of these are of the inner complex type, others may possibly be adsorption products. Alizarin itself in saturated alcoholic solution is sometimes used as a reagent for aluminium, but alizarinsulphonic acid is, perhaps, more convenient. A 0·1 per cent solution of sodium alizarinsulphonate in presence of dilute sodium hydroxide followed by dilute acetic acid produces a red colour or precipitate with very small traces of aluminium.

Quinalizarin (1: 2: 5: 8-tetrahydroxyanthraquinone) in dilute alkaline solution is a highly sensitive reagent for beryllium or magnesium. The reagent must be freshly prepared. Both elements produce an intense blue colour with this compound.

Alizarin and its derivatives are almost too sensitive for ordinary work. Blank tests are particularly necessary with these reagents.

11. *Benzidine*, $NH_2.C_6H_4.C_6H_4.NH_2$, is frequently used as a test for substances which have oxidizing properties. A dilute, 0·05 to 0·1 per cent, solution in acetic acid is commonly em-

ployed. Many oxidizing agents convert benzidine into blue oxidation products of a partially quinonoid character. An interesting adaptation of the molybdate reaction for phosphates to the micro-scale is carried out with the aid of benzidine (see p. 98). Two blue compounds are produced simultaneously, viz. the oxidation product of benzidine and the molybdenum blue resulting from the reduction of the ammonium phosphomolybdate.

12. *Guaiacol*, $C_6H_4{<}^{OH}_{OCH_3}$, is a highly sensitive reagent for nitrous acid. A saturated aqueous solution of beechwood creosote (which contains guaiacol) may be used. An intense orange-coloured para-nitroso derivative is produced.

13. *Hydrogen peroxide.* This reagent readily forms derivatives with the compounds of some metals which contain the metal in the acidic ion. Some of these derivatives are highly coloured. The deep blue perchromic acid, which is obtained by treating an acidified solution of a chromate or a dichromate with hydrogen peroxide, has been known for nearly a century. Coloured derivatives are also obtained with compounds of other metals, such as titanium and vanadium. The intense orange colour which acidified titanium compounds produce with hydrogen peroxide is highly sensitive, and well adapted to spot testing. It is possible to test qualitatively for a chromate and a vanadate in the same solution by acidifying it with dilute sulphuric acid, and adding hydrogen peroxide and ether. The blue perchromic acid is readily soluble in ether, and the ethereal solution floats above the aqueous layer which is coloured dark red by pervanadic acid.

14. *Oxalic acid.* The free acid, in addition to ammonium oxalate, is a useful reagent for characterizing cerium and other rare earths. Oxalic acid will precipitate cerous oxalate from a cerous salt containing free mineral acid, provided the concentration of the strong acid is not excessive. The oxalates of the true rare earths are insoluble in excess of oxalic acid or ammonium oxalate, whereas those of elements such as thorium form soluble oxalato-salts.

THE USE OF FILTER PAPER FOR
DROP REACTIONS

Feigl has drawn attention to the advantages which are to be derived from carrying out drop reactions on filter papers as compared with spot plates. The advantages arise in consequence of the porous properties of filter paper, which enable partial separations to be effected as the result of capillary and adsorption phenomena. By using filter papers, it is frequently possible to recognize two or even more constituents in a mixture by the use of a given reagent which if applied to a drop of the test solution on a spot plate would produce no definite result. Some examples of this may be quoted by way of illustration:

Detection of a thiocyanate in presence of a ferrocyanide. If ferric chloride is added to a solution of a mixture of the two salts, the colours of the ferric ferrocyanide precipitate and the ferric thiocyanate solution mutually obscure each other. If, however, a drop of the solution to be tested is placed on a filter paper, and a drop of ferric chloride is added, a separation is effected, resulting in the production of a central spot of ferric ferrocyanide which is blue surrounded by the red colour of ferric thiocyanate.

Detection of aluminium in presence of a ferric salt. A drop of the solution containing the two metals is placed on a filter paper. To this a drop of a solution of potassium ferrocyanide is added. The iron is precipitated as ferric ferrocyanide, resulting in the formation of a blue central spot. If an alcoholic solution of alizarin is now carefully dropped round the circumference of the liquid which has diffused away from the blue spot, and the filter paper exposed to the vapour of ammonia, a pink colour is produced due to the formation of a "lake" with the aluminium and alizarin.

Detection of aluminium in presence of uranium. Uranyl salts produce a blue "lake" with alizarin, whereas aluminium salts produce a pink "lake" with alizarin in the presence of ammonia. The two metals may be detected in the same solution by placing a few drops of an alcoholic solution of alizarin on a filter paper, which is then partially dried. On adding a drop of the solution to be tested, a blue spot is at once produced by reaction with the uranyl salt. If the filter paper is now exposed to the vapour of ammonia, the blue spot becomes surrounded by a pink ring, due to the diffusion of the aluminium salt in an outward direction.

It is very important to use filter paper as free as possible from metallic impurities for drop reactions. Swedish filter papers are treated with hydrofluoric and hydrochloric acids before being sent out, and are preferable to ordinary filter papers, which are liable to contain traces of silica, calcium, and magnesium oxide and even of ferric oxide. So-called "ashless" filter papers for gravimetric analysis are usually satisfactory for drop reactions, but it is frequently desirable to make blank tests on them in particular cases. This is particularly important in carrying out drop reactions when testing for traces of magnesium, or perhaps it would be more correct to say that the spot plate should be used for this element, since it has been shown by Feigl that para-nitrobenzene-azo-resorcinol, a reagent which is much used for magnesium, is adsorbed by magnesia-free filter paper with production of a blue colour.

Further information regarding the use of special reagents may be found in *Organic Reagents for Metals and for Certain Acid Radicals* by the Staff of the Research Laboratory of Hopkin and Williams, Ltd., Fourth Edition, London, 1943.

Part II
SPECIAL

Chapter III

REACTIONS OF THE METALS

(Highly sensitive reactions suitable for spot tests are indicated by asterisks ☼)

SILVER

Dilute hydrochloric acid produces a white precipitate of silver chloride, AgCl, which is darkened on exposure to light. Silver chloride is insoluble in dilute nitric acid, but is readily soluble in aqueous ammonia, due to the formation of complex cations, $\overset{+}{Ag(NH_3)_2}$. When the ammoniacal solution is acidified with nitric acid, the complex ion is decomposed and the chloride is re-precipitated. Silver chloride is also dissolved by solutions of potassium cyanide and sodium thiosulphate with formation of complex anions $\overset{-}{Ag(CN)_2}$ and $\overset{-}{Ag(S_2O_3)}$.

Potassium bromide and *potassium iodide* produce precipitates of the bromide and iodide respectively, the former being very pale yellow and the latter somewhat darker in colour. Both precipitates are insoluble in dilute nitric acid, but readily soluble in potassium cyanide and in sodium thiosulphate. The bromide is sparingly soluble in ammonia solution and the iodide is practically insoluble.

These reactions are employed in the gravimetric and volumetric determinations of the halogens and of silver.

Potassium thiocyanate produces a white precipitate of silver thiocyanate, AgSCN, insoluble in dilute nitric acid. This reaction is used for the volumetric determination of silver in acid solution.

Sodium hydroxide produces a brown precipitate of silver oxide, Ag_2O, insoluble in excess of the reagent. Silver oxide is also precipitated by ammonia, but is rapidly dissolved by excess of that reagent. Reducing agents such as glucose and alkaline tartrates readily precipitate the metal from ammoniacal solutions, and if the experiment is done carefully, the metal separates in mirror condition on the walls of the tube on warming gently.

✿ *Potassium chromate* precipitates silver chromate, Ag_2CrO_4, a crimson red compound soluble in strong acids, but insoluble in acetic acid.

Silver salts exert a marked *catalytic action* on some reactions which involve oxidation and reduction. Thus the oxidation of

✿ hydrochloric acid to chlorine by *ceric* salts is markedly catalysed by traces of silver salts. A dilute solution of ceric nitrate containing hydrochloric acid has an orange yellow colour, somewhat similar to that of a dichromate. Addition of a trace of a silver compound results in rapid reduction of the quadrivalent ceric salt to that of a colourless tervalent cerous salt. It is necessary to make a blank test under exactly parallel conditions. This test may be adapted for the purpose of recognizing the presence of silver in the precipitate produced by hydrochloric acid in systematic analysis. The precipitate is thoroughly washed with hot water, and a portion heated in a small crucible to volatilize any mercurous chloride. The residue after cooling is then treated with ceric nitrate as already described. Traces of silver also may be readily detected by their catalytic action on the oxidation of

✿ *chromic* salts to dichromates by persulphates. A drop of the solution to be tested is added to a drop of a dilute solution of chrome alum on a white tile. On adding a drop of a moderately concentrated solution of ammonium persulphate, the violet grey colour of the chromic salt disappears and is replaced by the orange yellow colour of the dichromate ion. A blank test must be carried out at the same time.

Solid silver compounds, mixed with *sodium carbonate*, when heated on charcoal in the reducing flame yield beads of the metal.

LEAD

Dilute hydrochloric acid precipitates lead chloride, $PbCl_2$, sparingly soluble in cold water, but readily in hot water from which the salt crystallizes on cooling.

Hydrogen sulphide precipitates black lead sulphide, PbS, from acid solutions. If a considerable excess of hydrochloric acid is present, a red salt, lead chlorosulphide, $PbSPbCl_2$, is produced. On dilution with much water the normal sulphide is obtained. Lead sulphide is dissolved by hot dilute nitric acid with formation of a solution of lead nitrate and separation of sulphur, but is insoluble in ammonium sulphide.

Potassium iodide produces a yellow precipitate of lead iodide, PbI_2, soluble in boiling water, from which it crystallizes in yellow spangles on cooling. Lead iodide is also soluble in sodium thiosulphate, complex ions being formed.

Potassium chromate precipitates yellow lead chromate, $PbCrO_4$, readily soluble in sodium hydroxide, soluble in mineral acids, but almost insoluble in acetic acid.

Dilute sulphuric acid precipitates white lead sulphate, $PbSO_4$. This salt is very sparingly soluble in water, still more insoluble in dilute alcohol, and is frequently used in the gravimetric estimation of lead. Lead sulphate is soluble in solutions of certain ammonium salts, such as ammonium tartrate, and acetate.

Sodium hydroxide precipitates the white hydroxide, $Pb(OH)_2$, readily soluble in excess of the reagent forming sodium plumbite, and in acids.

Alkaline oxidizing agents, such as hypochlorites, readily oxidize bivalent lead salts to the quadrivalent lead dioxide, PbO_2, which separates as a brown powder. This oxide is also produced by the anodic oxidation of lead immersed in a dilute solution of sulphuric acid (charging of a lead accumulator). A sensitive test for traces of lead, due to Trillat, depends on the production of an intensely blue oxidation product of tetramethyldiaminodiphenyl methane ✿ by the action of lead dioxide. Feigl (*Qualitative Analyse*), has recommended the test to be carried out as follows: Filter paper is moistened with an ammoniacal solution of 3 per cent hydrogen peroxide and a few drops of the solution to be tested are added. Lead dioxide is produced as a brown stain on the paper. After a few minutes a solution of the "tetra base" in acetic acid is dropped on the brown stain which soon develops a blue colour.

Metals which are more electropositive than lead displace the metal from its solutions. When *zinc* is immersed in a solution of lead acetate, preferably containing a little free acetic acid, a "tree" of crystalline lead (*arbor Saturni*) is produced. The appearance is very characteristic.

Solid lead compounds mixed with *sodium carbonate* and heated on charcoal in a reducing flame yield beads of the metal.

MERCURY

This metal gives rise to two series of salts, viz. mercurous salts corresponding to the doubly charged diatomic cation, $\overset{++}{Hg_2}$, which in many respects behaves as a univalent ion, and the mercuric salts which are bivalent and contain the cation $\overset{++}{Hg}$. The corresponding oxides are formulated as Hg_2O and HgO respectively.

The chemistry of mercury is in many respects anomalous, and this behaviour is to be attributed to the very limited ionization of its salts. The tendency, particularly of the mercuric salts, to form complex ions is well marked. The very small ionization of mercuric chloride, for example, is shown not only by osmotic effects and determinations of electrical conductivity, but can be demonstrated qualitatively as follows: If a little yellow mercuric oxide is shaken with water and phenolphthalein added, no colour appears in the solution. On the addition of sodium chloride the liquid shows an alkaline reaction, due to fixation of the chlorine ions by the mercury with consequent liberation of free hydroxyl ions. The salts of the oxyacids, on the other hand, show a pronounced tendency to undergo hydrolysis. When mercuric chloride is heated with concentrated sulphuric acid, no hydrogen chloride is evolved; ultimately the mercuric chloride volatilizes unchanged.

Mercurous salts.

Dilute hydrochloric acid precipitates white mercurous chloride, Hg_2Cl_2, insoluble in dilute nitric acid. Ammonia blackens the compound, probably as the result of the formation of a mixture of mercury amino-chloride, $Hg\begin{smallmatrix} \diagup NH_2 \\ \diagdown Cl \end{smallmatrix}$, and mercury. On account of its very sparing solubility, mercurous chloride is sometimes employed for estimating the metal gravimetrically. Mercurous chloride is dissolved by aqua regia being converted into mercuric chloride. Stannous chloride reduces it to mercury.

Sodium hydroxide precipitates black mercurous oxide, Hg_2O, insoluble in excess of the reagent.

Potassium chromate precipitates mercurous chromate,

$$Hg_2CrO_4,$$

as a reddish brown solid.

Potassium iodide produces a greenish precipitate of mercurous iodide, Hg_2I_2.

Mercuric salts.

Hydrogen sulphide produces a black precipitate of mercuric sulphide, HgS, insoluble in nitric acid, but soluble in aqua regia. If the gas is passed slowly through a solution of mercuric chloride, the precipitate which is first formed is white, becoming yellow then brown and finally black. These colour changes are due to the formation of mixed salts. Mercuric sulphide is insoluble in ammonium sulphide.

Sodium hydroxide produces a yellow precipitate of mercuric oxide, HgO.

Potassium iodide produces a yellow precipitate of mercuric iodide, HgI_2, which almost immediately becomes scarlet. Mercuric iodide is soluble in excess of either mercuric chloride or potassium iodide. In the latter reagent the mercury is present as a stable complex anion $\overset{-\ -}{HgI_4}$. A solution of potassium mercuric iodide containing excess of sodium or potassium hydroxide is a sensitive reagent for traces of ammonia (Nessler's reagent).

Stannous chloride readily reduces mercuric salts to the mercurous condition and ultimately to the metal. The precipitate which is first produced is white (calomel) and ultimately grey (finely divided mercury). Similar effects can be produced by other reducing agents, e.g. metals such as copper.

Ammonia produces a white precipitate of mercuric aminochloride, $Hg\begin{smallmatrix}\diagup NH_2 \\ \diagdown Cl\end{smallmatrix}$.

✿   A solution of *cobalt nitrate* mixed with *potassium thiocyanate* gives a deep blue crystalline precipitate of cobaltous mercuric thiocyanate, $Co[Hg(SCN)_4]$. This precipitate does not separate immediately from very dilute solutions, but crystallization is aided by scratching the interior of the vessel with a glass rod. The reaction is highly sensitive and adapted to the recognition of mercury on the micro-chemical scale.

✿   An alcoholic solution of *diphenylcarbazide* gives a blue violet colour with very small quantities of mercuric nitrate. This reaction may be carried out on filter paper. It is not specific for mercury, however, unless the absence of certain other substances, particularly chromates and molybdates, is assured.

Mercury has a remarkable capacity for uniting with other metals forming amalgams. With the alkali metals the reaction takes place with considerable evolution of heat. When small quantities of mercury are added to *aluminium*, rapid oxidation of the latter metal takes place with formation of feathery growths of the oxide. If a piece of aluminium foil is stroked with a glass rod, which has been dipped in a dilute solution of a mercuric salt, rapid formation of the oxide takes place where the rod has touched the metal. This test is very sensitive. The experiment may be varied by placing some aluminium turnings in hot water and adding a few drops of a dilute solution of the mercury salt; oxidation of the metal takes place almost immediately, accompanied with evolution of hydrogen.

The vapour pressure of mercury at the ordinary temperature though small is nevertheless quite definite. According to Stock, the vapour is highly poisonous; and care must therefore be taken to avoid prolonged exposure to it. Gold leaf exposed over mercury gradually becomes amalgamated, but the test, as ordinarily carried out, is not very sensitive.

THALLIUM

This metal gives rise to two series of salts, univalent thallous salts and tervalent thallic salts. The former are derived from the strongly basic oxide, Tl_2O, the latter from the very weakly basic oxide, Tl_2O_3. Thallium salts impart an intensely green colour to a non-luminous flame.

Thallous salts.

Dilute hydrochloric acid produces a white precipitate of thallous chloride, TlCl, sparingly soluble in cold water, but fairly readily soluble in hot water.

Potassium iodide precipitates yellow thallous iodide, TlI, very ☼ sparingly soluble in water, and used for determining the metal gravimetrically. Thallous iodide is insoluble in sodium thiosulphate, and is thereby readily distinguished from lead iodide.

Potassium chromate precipitates yellow thallous chromate, Tl_2CrO_4. Neither alkaline hydroxides nor ammonia produce any precipitate with thallous salts, since thallous hydroxide is readily soluble. Thallous hydroxide is as strong a base as the caustic alkalis.

Potassium thiocyanate produces a white sparingly soluble precipitate of thallous thiocyanate, TlSCN.

Thallic salts.

Sodium hydroxide or *ammonia* precipitates brown thallic hydroxide, $Tl(OH)_3$, insoluble in excess of either reagent, but readily soluble in hydrochloric acid.

Potassium iodide precipitates thallic iodide, TlI_3, as a black micro-crystalline powder. The compound behaves in some respects like thallous tri-iodide, but its *thallic* character predominates.

Thallic salts are easily reduced to the thallous condition, but powerful oxidizing agents, e.g. aqua regia or bromine, are required to effect the transformation of thallous salts into thallic salts. Alkaline potassium ferricyanide readily oxidizes solutions of thallous salts with separation of brown thallic oxide, Tl_2O_3, the ferricyanide being reduced to ferrocyanide:

$$Tl_2SO_4 + 4K_3Fe(CN)_6 + 6KOH$$
$$= Tl_2O_3 + 4K_4Fe(CN)_6 + 3H_2O + K_2SO_4.$$

Thallic oxide is a dark brown or black powder, the colour varying with the conditions of preparation. It is readily soluble in hydrochloric acid with formation of a solution of thallic chloride. Dilute sulphuric acid dissolves the oxide very slowly on heating, but the crystalline product which is produced in this way is always a basic or an acid sulphate: a normal sulphate is not obtained.

Both thallic sulphate and thallic nitrate are readily hydrolysed by water with separation of thallic oxide. The thallic halides are much less hydrolysed. Thallic chloride is extremely soluble in water, and if thallous chloride is dissolved in the hot solution, yellow crystals of thallous thallic chloride, $TlCl_3 3TlCl$, are obtained. Similar thallous thallic bromides can also be prepared, and also a dibromide, $TlBr_3 TlBr$, by dissolving thallous bromide in a concentrated solution of thallic bromide. The thallic halides readily form additive compounds with weak organic bases such as pyridine, which are sparingly soluble in water.

TUNGSTEN

The chief derivatives of tungsten are those of the acidic trioxide, WO_3, a bright yellow substance readily soluble in alkalis and in alkaline carbonates. Sodium tungstate, Na_2WO_4, and the other alkali tungstates are soluble in water, but all other tungstates are insoluble.

Dilute hydrochloric acid precipitates white tungstic acid, $WO_3(H_2O)_x$, in the cold and yellow tungstic acid if the solution is heated. Tungstic acid readily passes into colloidal solution. The difference between the varieties of tungstic acid seems to consist in differences in the water content. Tungstic acid is insoluble in excess of hydrochloric acid. Strong reducing agents, such as *stannous chloride*, in acid solution readily reduce tungstic ✿ acid to blue tungstic oxide, probably W_2O_5. This reaction is highly sensitive.

Hydrogen sulphide produces no precipitate in acid solutions, but if passed through a solution of an alkali tungstate, replacement of oxygen by sulphur takes place with formation of a thiotungstate. From such a solution, tungsten trisulphide, WS_3, may be precipitated on adding hydrochloric acid. The compound is soluble in ammonium sulphide.

Tungsten may be detected in the presence of molybdenum by an adaptation of the test for molybdenum with *hydrochloric acid,* ✿ *ammonium thiocyanate*, and *stannous chloride.* If a drop of the solution containing hydrochloric acid is placed on filter paper, a yellow spot due to tungstic acid at once appears. On adding ammonium thiocyanate and stannous chloride, the spot becomes blue due to the formation of the lower tungsten oxide, whilst molybdenum results in the production of the red complex thiocyanate. The red colour, however, disappears on adding concentrated hydrochloric acid, whereas the blue colour due to the tungstic oxide remains unchanged.

COPPER

Cupric salts.

Hydrogen sulphide precipitates black cupric sulphide, CuS, insoluble in dilute acids, but soluble in hot dilute nitric acid with separation of sulphur. Cupric sulphide is slightly soluble in yellow ammonium sulphide; and unless air is rigidly excluded, it is oxidized in the presence of water to cupric sulphate. It is also readily soluble in potassium cyanide with formation of a very stable complex anion, e.g. $\overline{Cu(CN)}_4$ or $Cu(\overline{CN})_2$.

Sodium hydroxide precipitates cupric hydroxide, $Cu(OH)_2$, as a blue gelatinous precipitate. This precipitate is readily soluble in solutions of certain organic compounds, rich in hydroxyl groups, such as tartrates, forming deep blue solutions in which the copper is present as a complex anion, e.g. $Cu\overline{C}_4\overline{H}_2O_6$, from which reducing agents such as glucose precipitate red cuprous oxide, Cu_2O, on heating. Fehling's solution, which is much used as a test for reducing agents, is prepared by mixing solutions of cupric sulphate, potassium sodium tartrate, and potassium hydroxide. When cupric hydroxide is heated in presence of water, or if sodium hydroxide is added to a boiling solution of a cupric salt, dehydration occurs, and black cupric oxide, CuO, separates. Cupric hydroxide is also readily soluble in *ammonia*, forming a deep blue solution in which the copper is present as a complex cation, $Cu(\overset{+}{N}\overset{+}{H}_3)_4$. This liquid has the property of dissolving cellulose.

Sodium carbonate precipitates a greenish blue precipitate of a basic carbonate, the composition of which is variable. If, however, a cupric salt is added to a saturated solution of potassium bicarbonate, a deep blue solution of potassium cupricarbonate is obtained. This liquid, Ost's solution, is sometimes used instead of Fehling's solution in testing for reducing agents, cuprous oxide being precipitated on heating.

Potassium ferrocyanide precipitates brown cupric ferrocyanide, ✳ $Cu_2[Fe(CN)_6]$, insoluble in dilute acids. This reaction is very sensitive.

Cuprous salts.

The cuprous salts are colourless, and only those which are insoluble in water, such as the halides and the thiocyanate, are stable. They are obtained by treating cupric salts with the appropriate precipitant in presence of a reducing agent.

Potassium iodide reduces solutions of cupric salts with separation of cuprous iodide, CuI, and liberation of iodine, which remains dissolved in excess of the reagent. If the experiment is carried out in presence of a suitable reducing agent, such as sulphurous acid or sodium thiosulphate, which removes the iodine, the cuprous iodide is obtained as a white precipitate. This reaction is employed for determining copper by a volumetric method.

Potassium thiocyanate, in presence of a reducing agent such as sulphurous acid, precipitates white cuprous thiocyanate, CuSCN. This reaction is used for the gravimetric determination of copper.

A solution of *salicylaldoxime* in dilute alcohol produces a ☼ yellowish green precipitate with cupric salts insoluble in acetic acid but soluble in mineral acids. The derivative is an internally complex compound of the formula $\left(C_6H_4 \underset{O-}{\overset{CH:NOH}{<}} \right)_2 Cu$ and is sometimes used for determining copper gravimetrically.

Metals, such as *zinc* or *iron*, readily displace copper from its solutions as a red deposit. Solid compounds of copper mixed with *sodium carbonate* yield red spangles of the metal when heated on charcoal in the reducing flame. The borax bead is pale blue, but if the bead is heated for some time in the reducing flame, especially in presence of a trace of tin, it turns red due to the formation of cuprous oxide. Copper compounds, especially the halides, impart an intense green colour to the non-luminous flame.

Certain reactions are catalysed by copper salts. An extremely sensitive test founded on this catalytic action is the following: Ferric salts are reduced to the ferrous condition by sodium thiosulphate in aqueous solution with formation of sodium tetrathionate:

$$2\overset{+++}{Fe} + 2\overset{--}{S_2O_3} = 2\overset{++}{Fe} + \overset{--}{S_4O_6},$$

a complex ion of a violet colour being produced as an intermediate compound. Minute traces of cupric salts accelerate the reduction enormously. The test is best carried out by mixing dilute solutions of ferric sulphate containing a little free sulphuric acid and sodium thiosulphate, and adding some potassium thiocyanate. A deep red solution is produced due to undissociated ferric thiocyanate. If the liquid thus obtained is divided into two parts, and a trace of a copper salt be added to one part, the red colour fades rapidly; the other portion remains red for a long time. The experiment is very suitable for a lecture-table demonstration. It is possible to demonstrate the presence of one part of copper in two and a half million parts of water by this means.

BISMUTH

Hydrogen sulphide precipitates the dark brown sulphide, Bi_2S_3, insoluble in ammonium sulphide. Hot dilute nitric acid dissolves the precipitate with separation of sulphur.

Sodium hydroxide or *ammonia* precipitates the white hydroxide, $Bi(OH)_3$, which is insoluble in excess of either reagent.

Bismuth salts are readily hydrolysed by *water* with formation of sparingly soluble basic salts. If a solution of bismuth chloride is diluted with much water a precipitate of bismuth oxychloride, BiOCl, is produced. This precipitate is insoluble in tartaric acid (distinction from antimony oxychloride).

Solutions of bismuth salts are reduced by *sodium stannite* to ✿ the metal. This test is carried out by preparing the reducing agent by adding sodium hydroxide to a solution of stannous chloride until the precipitate of stannous hydroxide is redissolved. The reagent is then added to the solution of the bismuth salt, when a black precipitate of the metal is produced at once. This test is extremely sensitive.

Potassium iodide produces a brown crystalline precipitate of the iodide, BiI_3. This precipitate is soluble in potassium iodide with formation of a complex anion, $\overline{Bi}I_4$. A solution of potassium bismuth iodide is sometimes used as a reagent for alkaloids.

✿   A solution of *cinchonine* in dilute nitric acid containing *potassium iodide* produces an intense orange colour.

✿   *Thiourea* in presence of dilute nitric acid produces an intense yellow colour.

Solid bismuth compounds mixed with *sodium carbonate* when heated on charcoal in the reducing flame yield beads of the metal. The beads are very brittle.

CADMIUM

Hydrogen sulphide precipitates yellow cadmium sulphide, CdS, insoluble in alkalis and in ammonium sulphide. Strong acids such as hydrochloric acid dissolve it readily; indeed, the precipitation of cadmium sulphide is very liable to be incomplete unless the acidity of the solution is maintained within somewhat narrow limits.

Sodium hydroxide and also *ammonia* precipitate the white hydroxide, $Cd(OH)_2$, insoluble in excess of the former reagent, but soluble in ammonia.

Potassium cyanide precipitates white cadmium cyanide, $Cd(CN)_2$, readily soluble in excess of the reagent, forming a complex anion $Cd(\overline{\overline{CN}})_4$. This anion is not very stable, and is readily decomposed by hydrogen sulphide with precipitation of cadmium sulphide. The separation of cadmium from copper is effected by taking advantage of the wide difference in the stability of the complex cyanides. When hydrogen sulphide is passed through a solution of a mixture of cadmium and cupric salts containing excess of potassium cyanide, cadmium sulphide only is precipitated.

An alcoholic solution of *diphenylcarbazide* produces in neutral ✿ solutions an intense red-violet precipitate or coloration. The test is best carried out on filter paper, a drop of the reagent followed by a drop of the solution being placed on the paper, which is then exposed to the vapour of ammonia. This reaction is not specific for cadmium; other metals, particularly mercury, behave similarly.

A solution of *mercuric chloride* containing excess of *ammonium* ✿ *thiocyanate* precipitates white crystalline cadmium mercuric thiocyanate, $Cd[Hg(SCN)_4]$. Zinc, however, gives a similar compound.

ARSENIC

Arsenious compounds.

Hydrogen sulphide produces a yellow precipitate of the sulphide, As_2S_3, in the presence of hydrochloric acid. In the absence of free acid or certain other electrolytes, the arsenious sulphide remains in colloidal solution. Arsenious sulphide is soluble in ammonium sulphide, in alkalis, and in alkaline carbonates, complex anions being formed. These salts are decomposed by hydrochloric acid with reprecipitation of arsenious sulphide.

Silver nitrate in presence of *sodium acetate* precipitates yellow silver arsenite,

$$Ag_3AsO_3, \text{ but its composition is variable,}$$

from solutions of neutral arsenites. The precipitate is readily soluble in acids and in ammonia.

Copper sulphate produces a green precipitate of a copper arsenite, the composition of which is variable. It is soluble in acids and in ammonia.

Copper placed in solutions of tervalent arsenic acidified with hydrochloric acid displaces arsenic from solution either as the element or as an arsenide of copper as a grey deposit on the metal (Reinsch's test). As antimony is liable to be precipitated under similar conditions, it is necessary to apply confirmatory tests for arsenic. This may be done by drying the piece of copper foil coated with the grey film with filter paper and heating it in a dry tube when the arsenic soon becomes oxidized to white crystalline arsenious oxide.

Oxidizing agents readily oxidize arseni*ous* compounds to the arseni*c* condition. Iodine in presence of sodium bicarbonate effects the oxidation quantitatively, and is used for estimating arsenic volumetrically.

Zinc and *dilute sulphuric* or *hydrochloric acid* reduce arsenious compounds to arsine, AsH_3. Highly sensitive tests for traces of arsenic depending upon the properties of arsine have been devised by Berzelius, Marsh and others. If a flame of burning hydrogen which contains minute traces of arsine is allowed to impinge on a cold porcelain surface, a brown mirror of arsenic is deposited on the porcelain. This film is at once dissolved by a solution of sodium hypochlorite. A solution of silver nitrate is readily reduced to the metal by the gas. The hydride of antimony, stibine, SbH_3, is produced under similar conditions, but is easily distinguished from that of arsenic by (1) the colour of the flames, that of arsenic being bluish while the flame of stibine is decidedly green, (2) the deposit of the element is black in the case of antimony and insoluble in hypochlorites. Moreover, the reaction with silver nitrate is different, a precipitate of silver antimonide being produced.

Aluminium and *sodium hydroxide* will generate arsine from a solution containing arsenic, but stibine is not produced under these conditions.

A very sensitive test for arsenic, due to Fleitmann, consists in allowing the hydrogen produced by the action of aluminium and sodium hydroxide in the presence of the substance to be tested for arsenic to come in contact with a filter paper moistened with a solution of silver nitrate, when the presence of a black stain on the paper indicates the presence of arsenic.

The importance of making blank tests with the reagents in carrying out Marsh's and Fleitmann's tests cannot be exaggerated.

Solid arsenic compounds are readily reduced to the element when heated on charcoal before the blowpipe. Arsenic is readily volatile and the vapour has a highly disagreeable odour resembling that of garlic.

Arsenic compounds (arsenates).

Hydrogen sulphide reduces arsenates acidified with hydrochloric acid very slowly with production of a precipitate of arsenious sulphide mixed with sulphur. Sulphur dioxide, however, reduces arsenates rapidly, after which sulphuretted hydrogen precipitates the arsenic readily as arsenious sulphide.

Silver nitrate produces a brick-red precipitate of silver arsenate, Ag_3AsO_4, readily soluble in mineral acids and in ammonia. Silver arsenate can, however, be precipitated quantitatively from a solution in dilute nitric acid if a sufficient excess of sodium acetate is added to the solution. This reaction has been adapted to the volumetric determination of arsenates.

Ammonium molybdate in presence of *nitric acid* gradually precipitates ammonium arsenomolybdate from a boiling solution as a yellow crystalline solid, somewhat similar to the corresponding phosphomolybdate.

Magnesia mixture (a mixture of magnesium sulphate containing excess of ammonium chloride and ammonium hydroxide) gives a white crystalline precipitate of magnesium ammonium arsenate, $MgNH_4AsO_4$, very similar to the corresponding phosphate. This reaction is useful for separating arsenates from other acidic radicals (except phosphates) and for distinguishing between arsenates and arsenites. When heated strongly, magnesium ammonium arsenate loses ammonia and water and is converted into the pyroarsenate, $Mg_2As_2O_7$.

In many respects, arsenates resemble phosphates very closely, but in addition are useful oxidizing agents.

ANTIMONY

Antimonious compounds.

Hydrogen sulphide produces in solutions acidified with hydrochloric acid an orange precipitate of the sulphide, Sb_2S_3, soluble in ammonium sulphide with formation of ammonium thioantimonite, from which it is reprecipitated by dilute acids. Antimonious sulphide is soluble in boiling concentrated hydrochloric acid with formation of antimonious chloride. When hydrogen sulphide is passed through a solution of an antimonious salt in the absence of free acid or other electrolyte the sulphide remains in colloidal solution.

Water readily hydrolyses solutions of antimonious salts with formation of sparingly soluble basic salts. Thus if a solution of antimonious chloride is diluted with much water, a precipitate of the oxychloride, $SbOCl$, is obtained. This precipitate is readily soluble in a solution of tartaric acid. It may be noted that a solution of potassium antimonyl tartrate, obtained by dissolving antimonious oxide in a solution of potassium bitartrate, is very much less hydrolysed than solutions of the inorganic antimonious salts. If a solution of tartar emetic is cautiously acidified with hydrochloric acid, a precipitate of the hydroxide, $Sb(OH)_3$, is produced, but this is readily soluble in excess of acid with formation of a solution of antimonious chloride.

Sodium hydroxide and *ammonia* precipitate the hydroxide or oxide, readily soluble in excess of the former reagent, but almost insoluble in excess of the latter.

Many metals, such as *zinc, iron, copper,* and *tin,* precipitate the metal from solution as a brown or black powder. Oxidizing agents must be absent. Some antimony hydride is liable to be formed.

Zinc and *dilute hydrochloric acid* readily generate antimony hydride, stibine, SbH_3, in solutions containing antimony. See p. 42 for Marsh's test for arsenic for differentiating between the two elements.

Solid antimony compounds mixed with *sodium carbonate* yield brittle beads of the metal when heated on charcoal before the blowpipe.

Antimonic compounds.

These compounds are obtained by treating the antimonious compounds with suitable oxidizing agents. Thus iodine, in presence of sodium bicarbonate, effects the transformation quantitatively, and the reaction is used in a volumetric method for determining the element. The antimonic compounds, which are derivatives of the oxide, Sb_2O_5, are wholly acidic in character. Antimonic acid is readily obtained by the hydrolysis of antimony pentachloride, and acid potassium pyroantimonate, $K_2H_2Sb_2O_7$, is sometimes used as a reagent for the detection of sodium, with which it forms a sparingly soluble sodium salt, $Na_2H_2Sb_2O_7$.

Hydrogen sulphide precipitates a mixture of antimonic sulphide, Sb_2S_5, antimonious sulphide, and sulphur, soluble in ammonium sulphide and in boiling concentrated hydrochloric acid with separation of sulphur.

Potassium iodide and *hydrochloric acid* reduce antimonic compounds to the antimonious condition with separation of iodine.

TIN
Stannous salts.

Hydrogen sulphide precipitates brown stannous sulphide, SnS, readily soluble in yellow ammonium sulphide as ammonium thiostannate, from which it is reprecipitated by dilute hydrochloric acid as yellow stannic sulphide, SnS_2. Stannous sulphide is soluble in concentrated hydrochloric acid.

Sodium hydroxide precipitates white stannous hydroxide, $Sn(OH)_2$, soluble in excess of the reagent with formation of sodium stannite. A solution of sodium stannite in excess of sodium hydroxide has very powerful reducing properties (see bismuth, p. 40).

Ammonium hydroxide or *carbonate* also precipitates the hydroxide, insoluble in excess of the reagent.

Mercuric chloride added to a solution of stannous chloride produces at first a white precipitate of mercurous chloride, Hg_2Cl_2, and afterwards a grey precipitate of mercury.

Oxidizing agents convert stannous salts quantitatively into the stannic condition. The reaction with iodine is employed for determining tin by a volumetric method.

Stannic salts.

Hydrogen sulphide precipitates yellow stannic sulphide, SnS_2, soluble in ammonium sulphide. Stannic sulphide is soluble in boiling concentrated hydrochloric acid with formation of a solution of stannic chloride.

Sodium hydroxide precipitates the white hydroxide, $Sn(OH)_4$, readily soluble in excess of the reagent with formation of sodium stannate.

Powerful reducing agents convert stannic salts into the stannous condition. Complete reduction is effected by boiling with *copper* in presence of *hydrochloric acid*. *Iron* reduces to the stannous condition at the ordinary temperature. With more electropositive metals, such as *zinc*, reduction proceeds as far as the metal, the tin being precipitated as a spongy mass.

Stannic salts undergo considerable hydrolysis in aqueous solution. Excess of hydrochloric acid must therefore be present to avoid precipitation of the hydroxide or a basic salt. With stannous salts, hydrolysis is less pronounced, but stannous salts are liable to oxidation by absorption of atmospheric oxygen with formation of oxychlorides. This oxidation may be avoided by adding granulated tin and excess of free hydrochloric acid to solutions of stannous chloride.

Both of the oxides of tin are amphoteric. The question of the nature of the various forms of stannic hydroxide (or stannic acid) has long been a debatable one. Some chemists have regarded the water as being adsorbed rather than chemically hydrated in these and in other hydroxides. Willstätter and others have investigated the effect of removing water from various metallic hydroxides with acetone. The results varied according to the mode of preparation and previous treatment of the materials. The general conclusion seems to be that in crystalline hydroxides the water is, in part at any rate, chemically combined, whereas in amorphous preparations it is chiefly adsorbed. The so-called metastannic acid, SnO_2xH_2O, is obtained by treating tin with concentrated nitric acid. It is insoluble in water, but when treated with concentrated hydrochloric acid it becomes soluble. The solution is, however, colloidal in character.

Solid tin compounds mixed with *sodium carbonate* yield beads of the metal when heated on charcoal before the blowpipe.

Traces of tin may be detected by utilizing the reducing properties of stannous chloride to bring about sensitive colour reactions. Two of such tests are as follows:

Ferric salts are readily reduced by stannous chloride to ferrous ☼ salts. No colour is produced when ferric salts and an alcoholic solution of dimethylglyoxime are mixed in presence of ammonia, but when a ferrous salt is formed by reduction, and intense red colour of the ferrous derivative of dimethylglyoxime is at once produced. The test is best carried out by mixing an acidified solution of *ferric sulphate* with a solution of the substance to be tested. Then after a minute, a little *tartaric acid* is added, followed by the *dimethylglyoxime* reagent and *ammonia*.

✿ *Ammonium phosphomolybdate* or *phosphomolybdic acid* is readily reduced by a trace of a stannous salt with production of the intense blue colour of molybdenum blue. The ammonium phosphomolybdate should be prepared in the usual way by the action of a considerable excess of ammonium molybdate on sodium phosphate acidified with nitric acid, but the precipitate must be thoroughly washed free from nitric acid. If phosphomolybdic acid is required, the ammonium phosphomolybdate should be dissolved in a little aqua regia, and then recrystallized from water.

MOLYBDENUM

The most important derivatives of this element are those of the acidic trioxide, MoO_3, a white substance which dissolves readily in alkalis to form molybdates. The element also exists in lower states of oxidation which are strongly coloured, some of which show resemblance to corresponding tungsten compounds. Permolybdates are also obtained by the action of hydrogen peroxide.

Hydrogen sulphide in presence of *hydrochloric acid* precipitates the dark brown trisulphide, MoS_3, with simultaneous reduction of the solution and production of a blue colour. Molybdenum trisulphide is soluble in yellow ammonium sulphide with formation of the thiomolybdate, and this solution is decomposed by acids with reprecipitation of the trisulphide.

A sensitive test for a molybdate consists in applying the familiar phosphomolybdate reaction in the reciprocal way. The solution is acidified with nitric acid and a trace of sodium phosphate is added. On warming the solution, the characteristic yellow precipitate of a phosphomolybdate is produced. The composition of ammonium phosphomolybdate is normally represented by the formula $(NH_4)_3PO_4$, $12MoO_3$, $6H_2O$. When heated, the residue has the composition $P_2O_5(MoO_3)_{24}$.

✿ A fine carmine red colour is produced by the action of *ammonium thiocyanate* in presence of *hydrochloric acid* and a *reducing agent*, such as *stannous chloride*. The colour is due to the formation of ammonium molybdothiocyanate,

$$(NH_4)_3[Mo(SCN)_6].$$

This reaction is well suited to filter paper. If iron is present, the red coloration due to ferric thiocyanate is removed by the action of the stannous chloride, which reduces the ferric salt to the ferrous condition.

Potassium xanthate in presence of dilute hydrochloric acid ✿ produces an intense red-violet colour. If more than traces of a molybdate are present, the compound separates in oily drops, which are readily soluble in organic solvents such as benzene, chloroform, or carbon disulphide. The formula

$$MoO_3[SC(SH)(OC_2H_5)]_2$$

has been suggested for the compound. The reaction is specific for molybdates.

A solution of *phenylhydrazine acetate* (containing excess of acetic acid) produces a red colour or precipitate. It is said that the colour is due to oxidation of the phenylhydrazine to a diazonium salt by the molybdate, and subsequent coupling of the diazonium salt with excess of the base.

IRON

Ferrous salts.

Ammonium sulphide precipitates black ferrous sulphide, FeS, which is insoluble in alkalis and alkaline sulphides, but readily soluble in acids with evolution of hydrogen sulphide.

Sodium hydroxide and *ammonia* precipitate ferrous hydroxide, $Fe(OH)_2$, which is greenish-white in colour but rapidly turns brown owing to absorption of oxygen, being converted into ferric hydroxide, $Fe(OH)_3$. In the presence of ammonium chloride, the precipitation of ferrous hydroxide by ammonia is impeded, and if the concentration of the ammonium chloride is considerable, it may fail altogether.

Potassium ferrocyanide produces a bluish white precipitate of potassium ferrous ferrocyanide, $K_2Fe[Fe(CN)_6]$, which rapidly becomes oxidized to a dark blue product, probably ferric ferrocyanide, $Fe_4[Fe(CN)_6]_3$ (Prussian blue).

✿   *Potassium ferricyanide* precipitates ferrous ferricyanide

$$Fe_3[Fe(CN)_6]_2$$

(Turnbull's blue), insoluble in acids.

Ammonium thiocyanate produces no colour with pure ferrous salts, but extremely small traces of ferric impurity result in the production of red ferric thiocyanate. This reagent is therefore used to ascertain the absence of ferric salts in ferrous salts. Some free acid should be present.

Solid iron compounds mixed with *sodium carbonate* yield the metal as magnetic powder when heated on charcoal before the blowpipe.

Solid iron compounds fused in a borax bead yield a dark green bead if the reducing flame is used. In the oxidizing flame, the bead assumes a reddish brown colour due to the formation of ferric oxide.

Ferric salts.

Alkaline hydroxides and *ammonia* precipitate ferric hydroxide, $Fe(OH)_3$, insoluble in excess of either reagent. The presence of ammonium chloride exercises no impeding influence, on account of the much weaker basic properties of ferric hydroxide as compared with ferrous hydroxide. Ferric hydroxide is readily soluble in acids, especially when it is freshly precipitated. When dried, it shrinks greatly in bulk and is then much more difficult to dissolve in acids.

Sodium acetate produces a dark red colour due to the formation of ferric acetate, $Fe(C_2H_3O_2)_3$, which is destroyed by mineral acids. On boiling a solution of ferric acetate, hydrolysis takes place and a precipitate of basic ferric acetate is produced, acetic acid escaping from the solution in the vapour. This reaction is of considerable importance in analysis, being used for the quantitative separation of iron and manganese, and for removing excess of iron in the ferric chloride method for the separation of phosphates. (See pp. 123 and 124.)

Potassium ferrocyanide produces a dark blue precipitate of ferric ferrocyanide, $Fe_4[Fe(CN)_6]_3$, insoluble in acids.

Potassium ferricyanide causes a darkening of the colour of a solution of a ferric salt but produces no precipitate. This reagent is therefore useful for testing ferric salts for the absence of ferrous salts, and is on this account used as an external indicator for titrating ferrous salts by potassium dichromate.

Ammonium thiocyanate produces a deep blood-red colour of ✻ ferric thiocyanate, $Fe(SCN)_3$, which is bleached by mercuric chloride, by phosphates, and also by fluorides due to the formation of the very stable ferrifluoride ion probably $\overline{\overline{\overline{FeF_6}}}$.

Oxidizing agents readily convert ferrous salts quantitatively into the ferric condition in acid solution. For volumetric analysis, the chief reagents are potassium permanganate, potassium dichromate, and ceric sulphate. The quantitative transformation of ferric salts into the ferrous condition may also be effected by many reducing agents, of which titanous chloride or titanous sulphate is particularly adapted to the requirements of volumetric analysis.

The ferric ion is apparently colourless, whereas the ferrous ion is very pale green. The solutions are, however, hydrolysed to some extent, with the result that they appear brown in colour. The addition of a little dilute sulphuric acid suppresses the hydrolysis and consequently the brown colour disappears. If, however, hydrochloric acid is added the liquid assumes a golden colour, which is not due to undissociated ferric chloride as was formerly supposed, but to the production of a complex ferrichloride ion, probably $\overline{\overline{\overline{FeCl_6}}}$. This ion is not stable, and the yellow colour disappears when the solution is diluted with much water.

CHROMIUM

Chromous salts.

These salts are derivatives of the bivalent chromous ion, which is bright blue in colour. They are obtained by reducing solutions of chromic salts or of chromates or dichromates with amalgamated zinc in presence of a large excess of hydrochloric acid. On pouring the resulting solutions into a concentrated solution of sodium acetate, chromous acetate is precipitated as a plum-red precipitate, which must be filtered with suction, washed with water saturated with carbon dioxide and finally with acetone. Solutions of other chromous salts may then be prepared by dissolving chromous acetate in the appropriate acid. They resemble ferrous salts in some respects, but are much more powerful reducing agents. Solutions of chromous chloride absorb oxygen rapidly, but have not been favoured for use in gas analysis on account of the tendency to liberate hydrogen. Solutions of silver, copper or mercury salts are at once reduced to the appropriate metal by chromous salts.

Chromic salts.

In the tervalent condition, the ion may be violet or green; and in general it may be stated that the colour of the hydrated chromic ion is usually violet, but as the tendency to form complex ions is considerable, green salts are very frequently produced. Solutions of most chromic salts are dichroic, i.e. the colour is different according as the solutions are viewed by reflected or by transmitted light.

Ammonia precipitates the green hydroxide, $Cr(OH)_3$, readily soluble in acids with formation of chromic salts, and in alkalis with formation of chromites. Precipitation by ammonia is liable to be incomplete unless the solution is boiled. Solutions of salts which develop alkaline hydrolysis or of weak bases also precipitate the hydroxide (compare aluminium, p. 54). The hydroxide may also be precipitated by Stock's reagent (a solution of potassium iodate and excess of potassium iodide).

Chromates and Dichromates.

These salts are derivatives of the trioxide, CrO_3, which is wholly acidic. The chromate anion, $\overset{--}{Cr}O_4$, is yellow, while the dichromate anion, $\overset{--}{Cr_2}O_7$, has an orange colour. Acids convert chromates into dichromates: $2\overset{+}{H} + 2\overset{--}{Cr}O_4 = H_2O + \overset{--}{Cr_2}O_7$, while alkalis reverse the process: $2\overset{-}{O}H + \overset{--}{Cr_2}O_7 = 2\overset{--}{Cr}O_4 + H_2O$. In presence of dilute hydrochloric or sulphuric acid, reducing agents, such as hydrogen sulphide or sulphur dioxide, convert dichromates into chromic salts. The conversion of chromic compounds into chromates may be effected by boiling with sodium peroxide, or alkaline hydrogen peroxide, or by fusion with sodium carbonate and a trace of a nitrate. Chromic salts can be oxidized in acid solution to dichromates by powerful oxidizing agents, ✿ one of the most efficient being ammonium persulphate, especially if a small quantity of a silver salt be added. This reaction forms the basis of a volumetric method for determining chromium, the excess of persulphate being destroyed by boiling.

Hydrogen peroxide added to an acidified solution of a chromate or a dichromate produces an intense indigo-blue colour due to the formation of a perchromic acid, $HCrO_5$. The compound is very unstable in aqueous solution, and undergoes rapid reduction to a chromic salt with evolution of oxygen. In ethereal solution perchromic acid is much more stable. This reaction furnishes a sensitive test for chromates, and reciprocally for hydrogen peroxide. In using the reaction as a test for hydrogen peroxide, however, it is necessary to carry out a blank test on the ether to ascertain that the solvent is free from organic peroxides.

Barium chloride produces a yellow precipitate of barium chromate, $BaCrO_4$, soluble in strong acids, but insoluble in acetic acid. This compound is sometimes used for separating barium from strontium and calcium.

Solutions of heavy metals, such as *silver nitrate, lead acetate,* and *mercurous nitrate,* give precipitates of the corresponding chromates, that of the first being dark red, the second yellow, and the third terracotta in colour. (See the reactions of these metals.) These precipitates are soluble in dilute nitric acid, but practically insoluble in acetic acid.

Concentrated sulphuric acid added to a solution of an alkaline chromate or dichromate precipitates chromium trioxide:

$$Na_2Cr_2O_7 + H_2SO_4 = Na_2SO_4 + H_2O + 2CrO_3.$$

Chromium trioxide or chromic anhydride is a crimson-red crystalline solid, very soluble in water. To obtain the compound in anhydrous condition, the crystals are filtered through asbestos or glass wool, washed with concentrated nitric acid and allowed to dry in a desiccator. The compound is exceedingly hygroscopic, and is an extremely powerful oxidizing agent. With some organic compounds, such as alcohol, the reaction is so violent that inflammation takes place. When dissolved in water, a mixture of chromic and dichromic acids is produced. In solution there is an equilibrium between the two acids:

$$2H_2CrO_4 \rightleftharpoons H_2Cr_2O_7 + H_2O,$$

dilution favouring the production of chromic acid.

Solid chromium compounds when fused in a borax bead give a fine green colour.

ALUMINIUM

Ammonia precipitates white aluminium hydroxide, $Al(OH)_3$, readily soluble in acids and in alkalis. The hydroxide is liable to remain in the hydrosol condition, from which it separates as the hydrogel by boiling or by adding electrolytes. When ammonia is used to effect the precipitation, ammonium chloride should always be present to avoid the simultaneous precipitation of more strongly basic hydroxides.

Sodium hydroxide precipitates the hydroxide, very readily soluble in excess of the reagent with formation of a solution of sodium aluminate, $NaAlO_2$, in which the aluminium is present in the anion. If carbon dioxide be passed through the solution, the hydroxide is reprecipitated:

$$NaAlO_2 + CO_2 + 2H_2O = Al(OH)_3 + NaHCO_3.$$

Other sources of hydroxyl ions, such as salts which show alkaline hydrolysis like ammonium sulphide, sodium carbonate, sodium thiosulphate and sodium nitrite, or weak bases such as aniline or phenylhydrazine, readily precipitate the hydroxide. The use of these precipitants is to be recommended in preference to ammonia when it is required to precipitate aluminium hydroxide quantitatively in gravimetric analysis.

Potassium iodate in presence of excess of *potassium iodide* precipitates the hydroxide with separation of iodine. Aluminium salts exhibit strong acid hydrolysis on account of the feebly basic character of aluminium hydroxide, and the acid resulting from hydrolysis liberates iodine from the iodate-iodide mixture:

$$3H_2O + 2AlCl_3 + KIO_3 + 5KI = 2Al(OH)_3 + 6KCl + 3I_2.$$

By adding sufficient *sodium thiosulphate*, the whole of the iodine can be removed from solution and the precipitation of the aluminium hydroxide effected quantitatively. This reaction has been applied by Stock to the quantitative separation of aluminium from more strongly basic metals.

Solid aluminium compounds when strongly heated on charcoal before the blow-pipe yield an infusible residue of the oxide. When this residue is moistened with a solution of cobalt nitrate and heated a second time, it assumes a blue colour (Thénard's blue), sometimes described as cobalt aluminate; it may, however, be a solid solution of the two oxides. A better method of carrying out this test consists in soaking a filter paper in a solution of the substance previously mixed with cobalt nitrate, or better still, with potassium cobalticyanide, and calcining the paper. The presence of aluminium is easily seen by examining the colour of the ash.

Aluminium salts produce an intensely coloured red "lake" with *alizarinsulphonic acid*. The test is best carried out by ☆ placing a drop of the solution containing excess of sodium hydroxide, so as to have a solution of sodium aluminate, on a spot plate, adding a few drops of an aqueous solution of sodium alizarinsulphonate, followed by drops of dilute acetic acid. In the presence of aluminium, the deep violet colour which is produced by the action of the alkali on the sodium alizarinsulphonate is replaced by a bright red precipitate or colour. As the reaction is highly sensitive, it is desirable to make a blank test on the sodium hydroxide solution to prevent the possibility of error arising from the presence of aluminium in that reagent.

The "lake" is probably an internally complex salt of aluminium, for which the following formula has been suggested:

$$Al/3$$

O   O

OH

SO_3H

O

Alizarin in alcoholic solution and in presence of ammonia produces a red "lake" somewhat similar in appearance to that obtained with sodium alizarinsulphonate. The test is most con- ✿ veniently carried out by placing a drop of the solution in dilute hydrochloric acid on a filter paper, adding a few drops of an alcoholic solution of alizarin, and then exposing the paper to the vapour of ammonia. In all cases, it is highly desirable to make a blank test under identical conditions, because a violet colour is produced by the action of ammonia on filter paper which has been moistened with alizarin dissolved in alcohol.

✿ *Morin*, $C_{15}H_{10}O_7$, $2H_2O$, dissolved in methyl alcohol produces an intense green fluorescence with aluminium salts in aqueous or dilute acetic acid solution, consisting of a colloidal solution of the compound $Al(C_{15}H_9O_7)_3$.

TITANIUM

This element exists in more than one state of oxidation of which the derivatives of the tervalent ion, which have a fine violet colour, and those of the very weakly basic quadrivalent ion, which are colourless, are the most familiar. In addition, pertitanates, corresponding to the oxide TiO_3, are known. The titanous salts are strong reducing agents, and have considerable application in volumetric analysis.

Reactions of titanic compounds. These compounds of quad-rivalent titanium may be obtained from the dioxide, TiO_2. The oxide is very sparingly soluble in dilute acids, but more readily so in concentrated sulphuric acid. On dilution with water, a solution of the reagent, frequently termed titanic acid, is obtained.

Alkalis, or indeed practically any source of hydroxyl ions, precipitate the hydroxide as a white voluminous precipitate. The precipitate is soluble in acids, but practically insoluble in alkalis. Modifications of hydrated titanium dioxide, known as orthotitanic and metatitanic acid, have been described, the differences in properties being connected with the state of aggregation and the amount of water combined or adsorbed.

Powerful *reducing agents*, such as *zinc* in acid solution, reduce solutions of quadrivalent titanium to the tervalent condition, the liquid becoming violet.

Hydrogen peroxide produces an intense yellow or brown colour ✿ according to the amount of titanium in solution. The colour is said to be that of free peroxo-disulphato-titanic acid, in which the titanium is present as a complex anion, $[TiO_2(SO_4)_2]^{--}$. The production of the colour is liable to interference in the presence of substances such as chromates, vanadates or other substances which yield coloured derivatives of hydrogen peroxide. On the other hand, the reaction is an extremely valuable means of testing for hydrogen peroxide.

Colour reactions, particularly with certain phenols, have been ✿ recommended for testing for traces of titanium. Thus *pyrocatechol* produces in solutions, feebly acidified with *dilute sulphuric acid*, an intense yellow colour. This reaction is suitably carried out on filter paper.

A sensitive colour reaction, some twenty times as sensitive as ✿ the hydrogen peroxide reaction, is given with *dihydroxymaleic acid*, $\begin{array}{l} C(OH)COOH \\ \| \\ C(OH)COOH \end{array}$ (Fenton). This colour reaction is not interfered with by vanadium compounds. A colorimetric method of determining titanium and vanadium salts in dilute sulphuric acid solution by the use of Fenton's acid for the titanium and hydrogen peroxide for the vanadium has been described by Mellor (*Trans. Cer. Soc.* 1913, **12**, p. 33).

BERYLLIUM

This element resembles aluminium and also magnesium. Indeed, it may be described as intermediate in properties between these two elements. It is less basic than magnesium, but decidedly more so than aluminium. Most of its analytical reactions are similar to those of aluminium.

Ammonia precipitates the hydroxide, $Be(OH)_2$, as a white flocculent precipitate, very similar to aluminium hydroxide in appearance. It is soluble in acids and also in alkaline hydroxides with formation of beryllates. These compounds are, however, more readily hydrolysed than aluminates. The precipitation of beryllium hydroxide is impeded by the presence of substances which are rich in hydroxyl groups, such as tartrates.

Alkaline carbonates precipitate a basic beryllium carbonate of indefinite composition. The normal carbonate is not obtained on account of the weakly basic properties of beryllium. The precipitate is soluble in ammonium carbonate.

The separation of beryllium from aluminium is somewhat difficult. One method for effecting this is to add excess of a solution of *sodium bicarbonate* to the solution. This results in the formation of the hydroxides of both metals. Beryllium hydroxide readily dissolves in excess of the hot reagent, but aluminium hydroxide is insoluble. Methods depending on the use of *tannin* have been described by Schoeller (*The Analytical Chemistry of Tantalum and Niobium*, 1937).

Traces of beryllium may be detected by the use of *quinalizarin*, ✿ 1 : 2 : 5 : 8-tetrahydroxyanthraquinone. A very dilute alkaline solution of this dyestuff (0·05 g. in 100 cc. of normal sodium hydroxide), which must be freshly prepared, has an intense violet colour. The presence of a trace of a beryllium salt results in the production of a deep cornflower blue colour. It must be noted that a similarly coloured derivative is formed with magnesium salts.

CERIUM

Cerium is the only metal belonging to the rare earths which exists in two well-defined states of oxidation. The tervalent cerous salts, which are practically colourless, are derivatives of the strongly basic oxide, Ce_2O_3. The quadrivalent ceric salts, derived from the oxide, CeO_2, are orange coloured and are powerful oxidizing agents. The ceric salts are considerably hydrolysed in solution on account of the weakly basic character of the oxide, and the tendency to form complex ions is well marked.

Cerous salts.

Ammonia or *sodium hydroxide* precipitates the white hydroxide, $Ce(OH)_3$, insoluble in excess of either reagent.

Ammonia in the presence of *hydrogen peroxide* produces a ✻ precipitate of the yellowish brown perhydroxide, $Ce(OH)_3(O_2H)$. This reaction is highly sensitive and characteristic for cerium. It is, however, not directly applicable in the presence of iron, since the colour of ferric hydroxide is similar to that of ceric perhydroxide. If, however, excess of an alkaline tartrate is added to the solution, the precipitation of ferric hydroxide is prevented in consequence of the formation of complex ferri-tartrate ions, and it is then possible to recognize the presence of cerium.

Alkaline carbonates precipitate white cerous carbonate, $Ce_2(CO_3)_3$, insoluble in excess of sodium carbonate, and very nearly insoluble in excess of ammonium carbonate.

Oxalic acid or a *soluble oxalate* precipitates white cerous oxalate, $Ce_2(C_2O_4)_3$, insoluble in excess of the precipitants and in dilute acids, and very sparingly soluble in dilute hydrochloric acid. The oxalates of the rare earths are exceptional in this respect, and this behaviour is of importance in effecting their separation from other substances. The insolubility of the oxalates of the rare earths in excess of ammonium oxalate enables them to be separated from zirconium and thorium, the oxalates of which are soluble in excess of that reagent.

Ceric salts.

When cerous carbonate or cerous oxalate is ignited in an open crucible, the residue consists of ceric oxide. In the absence of traces of other substances, particularly of other rare earths, the colour of ceric oxide is pale yellow, but as usually obtained the oxide is of a cinnamon brown colour.

Concentrated *hydrochloric acid* dissolves ceric oxide with formation of a dark reddish brown solution of ceric chloride, which readily loses chlorine on warming with formation of cerous chloride.

Concentrated sulphuric acid gradually converts ceric oxide into the sulphate without visibly passing into solution, and without loss of oxygen. The product is, however, readily soluble in water, and the solution is very stable. A solution of ceric sulphate is a useful oxidizing agent for use in volumetric analysis.

Reducing agents readily convert ceric salts into the colourless cerous salts. *Hydrogen peroxide* causes reduction with simultaneous evolution of oxygen.

THORIUM

This element is quadrivalent, and exists in only one state of oxidation. All the salts, except such as are derived from coloured anions, are colourless. The hydroxide, $Th(OH)_4$, possesses purely basic properties, nevertheless the salts derived from strong acids exhibit a certain amount of acid hydrolysis. A well-defined tendency to form complex ions is shown by certain compounds, particularly the oxalate and the carbonate.

Ammonia precipitates the hydroxide, $Th(OH)_4$, as a white gelatinous precipitate, insoluble in excess of the reagent, but readily soluble in acids. It is, however, soluble in solutions of alkaline carbonates. Other sources of hydroxyl ions, such as ammonium sulphide, produce the same precipitate.

Alkaline carbonates precipitate the white basic carbonate, readily soluble in excess of these reagents, with formation of salts of the type of $Na_6[Th(CO_3)_5]$.

Oxalic acid or *ammonium oxalate* precipitates white thorium oxalate, $Th(C_2O_4)_2$, insoluble in excess of oxalic acid, but soluble in excess of ammonium oxalate. The complex salts which are formed with excess of ammonium oxalate are of the type $(NH_4)_4[Th(C_2O_4)_4]$.

Hydrogen peroxide precipitates a peroxide as a white gelatinous product, in which the atomic ratio of thorium to active oxygen is 2 to 3.

ZIRCONIUM

This element is quadrivalent, the compounds being derived from the weakly basic oxide, ZrO_2. The salts are derived chiefly from the bivalent zirconyl ion, $\overset{+\;+}{ZrO}$, and are colourless, unless the anion happens to be coloured. The tendency to form complex ions is well defined.

Ammonia or the *alkali hydroxides* precipitate the white hydroxide, which is gelatinous and readily forms colloidal solutions. It is soluble in acids.

Ammonium carbonate precipitates a basic carbonate, soluble in excess of the reagent, but rather sparingly soluble in excess of sodium carbonate.

Oxalic acid precipitates the oxalate, $ZrO(C_2O_4)$, soluble in excess of the reagent and in ammonium oxalate, due to the formation of complex zirconyl oxalates.

Hydrogen peroxide added to a neutral solution of a zirconium salt produces a white gelatinous precipitate of a perzirconic acid, similar in constitution to the derivative obtained by the action of the reagent upon quadrivalent titanium compounds.

Turmeric paper dipped into a solution of a zirconium compound, acidified with dilute hydrochloric or sulphuric acid and dried, assumes a reddish yellow colour. Titanium compounds interfere if present in the quadrivalent condition. They must therefore be reduced to the tervalent condition by zinc and dilute hydrochloric or sulphuric acid before the test is applied.

Sodium alizarinsulphonate added to a solution of a zirconium salt, in the presence of dilute hydrochloric acid, results in the production of an intense reddish violet colour. If a solution of an alkali fluoride is added the colour is changed to yellow in consequence of the formation of a stable anion, $\overline{\overline{ZrF_6}}$.

URANIUM

Uranyl salts.

These salts are of the type UO_2X_2, in which the uranyl cation is bivalent and X denotes a univalent acidic residue. The uranium in these compounds is sexivalent, corresponding to the amphoteric oxide UO_3. The uranyl salts possess a greenish yellow colour having a marked fluorescence. The nitrate and acetate are very soluble in water. When ignited, a residue of the black oxide, U_3O_8, which behaves as $UO_2(UO_3)_2$, is ultimately obtained.

Ammonia precipitates ammonium di-uranate, $(NH_4)_2U_2O_7$, as a yellowish powder, insoluble in excess of the reagent, but readily soluble in ammonium carbonate with formation of a complex salt:

$$2UO_2(NO_3)_2 + 6NH_4OH = (NH_4)_2U_2O_7 + 4NH_4NO_3 + 3H_2O.$$

Ammonium carbonate precipitates the carbonate, UO_2CO_3, which dissolves in excess of the reagent, forming a clear yellow solution containing the complex salt $(NH_4)_4[UO_2(CO_3)_3]$. This behaviour enables uranium to be separated from the metals of the iron group, which are precipitated as hydroxides.

Ammonium sulphide precipitates brown uranyl sulphide, UO_2S, insoluble in excess of the reagent, but readily soluble in acids including acetic acid.

✿ *Potassium ferrocyanide* precipitates the dark brown ferrocyanide, $(UO_2)_2[Fe(CN)_6]$.

Reducing agents convert uranyl salts into the quadrivalent uranous salts, which have a green colour.

Uranous salts.

These salts possess powerful reducing properties. *Oxidizing agents* transform them into uranyl salts. The reaction with acidified potassium permanganate is quantitative, and is used in volumetric work.

NICKEL

Ammonium sulphide precipitates black nickelous sulphide, NiS. This compound is appreciably soluble in yellow ammonium sulphide, forming a dark brown solution, but is nearly insoluble in the colourless reagent. Nickelous sulphide is practically insoluble in dilute hydrochloric acid, but is readily dissolved by aqua regia, forming a solution of nickelous chloride. Hydrogen sulphide does not in general precipitate nickelous sulphide from a nickelous salt, unless a "buffer" such as sodium acetate is present in the solution.

Sodium hydroxide precipitates green nickelous hydroxide, insoluble in excess of the reagent, but soluble in ammonium salts. *Ammonia* produces the same precipitate, readily soluble in excess, forming a blue solution.

Potassium cyanide precipitates greenish yellow nickelous cyanide, readily soluble in excess of the reagent, forming the complex salt, $K_2Ni(CN)_4$. Unlike the corresponding cobalto-cyanide, this compound is not oxidized to a complex nickel salt, but on account of its lower stability, alkaline oxidizing agents, such as sodium hypochlorite or hypobromite, effect its transformation into black nickelic hydroxide, $Ni(OH)_3$, on heating gently, thus:

$$2Ni(CN)_2 + NaOBr + 5H_2O = 2Ni(OH)_3 + NaBr + 4HCN.$$

Dimethyl glyoxime dissolved in alcohol produces, in the presence of ammonia, a scarlet precipitate of the nickel derivative of the reagent:

$$NiSO_4 + 2NH_3 + \begin{matrix} 2CH_3.C:NOH \\ | \\ CH_3.C:NOH \end{matrix} = \begin{matrix} CH_3.C:NOH \\ | \\ CH_3.C:NO \end{matrix} \diagdown Ni \diagup \begin{matrix} ON:C.CH_3 \\ | \\ HON:C.CH_3 \end{matrix} + (NH_4)_2SO_4.$$

This compound is used for determining nickel gravimetrically. Relatively few other metals interfere with the reaction, but mention must be made of ferrous iron and of palladium, both of which yield derivatives with the reagent.

Cobalt

Ammonium sulphide precipitates black cobaltous sulphide, CoS, insoluble in excess of the reagent and in dilute hydrochloric acid. Cobaltous sulphide is, however, not precipitated by hydrogen sulphide from solutions of cobalt salts unless a substance such as sodium acetate is present which diminishes the effective acidity of the solution. Aqua regia dissolves the precipitate readily, a solution of cobaltous chloride being produced.

Sodium hydroxide precipitates a blue basic salt at the ordinary temperature, but on heating the solution the pink hydroxide, $Co(OH)_2$, is obtained. On exposure to the air, oxidation to the brown cobaltic hydroxide gradually takes place. Alkaline oxidizing agents, such as sodium hypochlorite or hypobromite, precipitate the black dioxide, $CoO_2 x H_2O$.

Ammonia in presence of *ammonium chloride* produces no precipitate, but the solution thus obtained undergoes oxidation on exposure to the air with formation of various cobaltammines, e.g. hexammine cobaltic chloride, $[Co(NH_3)_6]Cl_3$, aquo-pentammine cobaltic chloride, $[Co(NH_3)_5H_2O]Cl_3$, and chloro-pentammine cobaltic chloride, $[ClCo(NH_3)_5]Cl_2$.

✱ *Ammonium thiocyanate* in presence of *acetone* produces an intense blue colour, due to the formation of a solution of ammonium cobaltous thiocyanate, $(NH_4)_2[Co(SCN)_4]$.

✱ *Ammonium mercurithiocyanate* precipitates blue cobaltous mercurithiocyanate, $Co[Hg(SCN)_4]$, especially on scratching the vessel with a glass rod.

Potassium cyanide precipitates cobaltous cyanide, $Co(CN)_2$, readily soluble in excess of the reagent with formation of potassium cobaltocyanide. The solution readily undergoes oxidation, particularly with the aid of an alkaline oxidizing agent, such as sodium hypochlorite or hypobromite, with formation of the very stable cobalticyanide, $K_3Co(CN)_6$.

✱ *α-Nitroso-β-naphthol* dissolved in acetic acid when added to a neutral solution of a cobaltous salt, or slightly acidified with hydrochloric acid, precipitates the cobalt quantitatively as the orange coloured cobalti-nitroso naphthol, $(C_{10}H_6O_2N)_3Co$.

Potassium nitrite (concentrated solution) in presence of acetic acid produces a yellow crystalline precipitate of potassium cobaltinitrite, $K_3Co(NO_2)_6$, with simultaneous evolution of nitric oxide. Presumably cobaltous nitrite is first produced, which is subsequently oxidized to cobaltic nitrite by the nitrous acid, which in its turn is reduced to nitric oxide. The cobaltic nitrite then forms a complex salt with excess of potassium nitrite containing the stable complex anion, $Co(\overline{NO_2})_6$.

The three last-named reagents, particularly α-nitroso-β-naphthol and potassium nitrite, are used for effecting the quantitative separation of cobalt from nickel.

Cobalt compounds when fused in a borax bead impart an intense blue colour to the bead. The colour is the same whether the fusion is effected in the oxidizing or in the reducing flame.

The colour of aqueous solutions of cobaltous salts is red, which is said to be that of the hydrated cobaltous ion. If concentrated hydrochloric acid be added to a solution of cobaltous chloride, the colour changes from red to blue, but reverts to red if the solution is diluted. These changes may be discussed in terms of hydration and dehydration, as it is well known that the higher cobaltous hydrates are red, whereas the monohydrate and the anhydrous salt are blue. In solvents such as acetone or acetonitrile, cobaltous chloride is blue, but in methyl alcohol the colour is red. It may be added that solutions of nickel salts are green, the colour being complementary to that of cobalt. A solution of mixture of a cobalt and a nickel salt may be red, green or colourless according to the relative proportions of the two constituents.

Cobalt and nickel are very similar in properties and reactions, but it may be remarked that the tendency to form stable coordination compounds is more strongly developed in cobalt than in nickel. This behaviour is in accordance with the atomic numbers, Co 27, and Ni 28, which follow Fe 26.

Cobalt may be detected in the presence of iron with the aid of the thiocyanate reaction. If a solution of *ammonium thiocyanate* is added to a solution containing a ferric salt and a cobaltous salt, the deep red colour of ferric thiocyanate is at once produced. On adding a solution of *sodium fluoride*, the red colour is discharged on account of the formation of the very stable ferrifluoride ion $\overline{\overline{\overline{FeF_6}}}$. If *acetone* be added to the solution, the blue colour due to the production of the complex salt

$$(NH_4)_2[Co(SCN)_4]$$

becomes apparent.

MANGANESE

Manganous salts.

These salts are derivatives of bivalent manganese, the ion being faintly pink, or even colourless according to some authorities. They are the most stable compounds of the element.

Ammonium sulphide precipitates pink manganous sulphide, MnS, insoluble in excess of the reagent, but readily soluble in acids, including acetic acid.

Sodium hydroxide and also *ammonia* precipitate the white hydroxide, $Mn(OH)_2$, insoluble in excess of either reagent. The precipitate readily absorbs oxygen with formation of the hydrated sesquioxide or dioxide, having a brown colour. In the presence of ammonium chloride, ammonia does not precipitate manganous hydroxide; but after a short time, the solution undergoes oxidation with separation of the higher hydroxide. The separation of ferric iron from manganese by ammonia in presence of ammonium chloride is therefore always incomplete; the precipitate invariably contains some manganese.

Alkaline oxidizing agents, such as *sodium hypochlorite*, readily ✿ oxidize solutions of manganous salts to hydrated manganese dioxide. A sensitive test depending upon the use of ammoniacal silver nitrate as the oxidizing agent may be carried out as a drop reaction on filter paper. In the presence of manganese, a black stain of silver together with manganese dioxide is produced, thus:

$$\overset{++}{Mn} + 2Ag[(\overset{+}{NH_3})_2] + 4\overset{-}{O}H = 2Ag + MnO_2 + 4NH_3 + 2H_2O.$$

This reaction may be adapted as a test for ammonia by mixing solutions of a manganous salt and silver nitrate, placing a drop of the solution on filter paper, and exposing it to the gas which it is desired to test.

Fused with *sodium carbonate* and a trace of *potassium nitrate*, manganese compounds yield a green melt of sodium manganate.

A sensitive reaction for traces of manganese depends upon the production of an intense blue coloured oxidation product of ✿ *benzidine*. When a manganous salt is precipitated by an alkaline hydroxide or by ammonia, the resulting manganous hydroxide undergoes rapid oxidation to the hydrated dioxide on exposure to the air. If this reaction is caused to take place in presence of a solution of benzidine in dilute acetic acid, the blue quinonoid oxidation product is at once produced. This reaction cannot be used as a test for manganese in the presence of other oxidizing agents which are capable of oxidizing benzidine, such as chromates, ferricyanides, and cerium salts. It is, however, of value in testing for traces of manganese in the presence of iron, because ferric hydroxide does not oxidize benzidine under the conditions required for the test. The best way of carrying out the test is to add a sufficient quantity of an alkaline tartrate, such as Rochelle salt, to prevent the precipitation of ferric hydroxide when ammonia or dilute sodium hydroxide is added to the solution. A drop of the solution is then placed upon filter paper, followed by a drop of a dilute solution of benzidine in dilute acetic acid. The presence of a very small quantity of manganese is sufficient to result in the production of a blue colour.

When fused in a borax bead, manganese compounds impart a violet colour to it if the oxidizing flame is used. In the reducing flame, the bead becomes colourless.

Derivatives of tervalent and quadrivalent manganese.

Unstable compounds in which manganese functions as a tervalent or quadrivalent element can be prepared by dissolving the sesquioxide, Mn_2O_3, manganosomanganic oxide, Mn_3O_4, or manganese dioxide in the appropriate acid. The solutions, which are dark red or violet in colour, are very unstable unless a considerable excess of acid is present. Dilution with water results in the separation of the higher hydrated oxide and the formation of a solution of a manganous salt. Thus manganic sulphate undergoes hydrolysis, presumably as follows:

$$Mn_2(SO_4)_3 + 2H_2O = MnO_2 + MnSO_4 + 2H_2SO_4.$$

Greater stability is found in complex derivatives. Thus manganic sulphate forms a fairly stable alum with caesium sulphate.

Manganates and permanganates.

Manganates are derivatives of the green bivalent anion $\overset{=}{MnO_4}$, while permanganates are derived from the deep violet anion $\overset{-}{MnO_4}$, which is univalent. Aqueous solutions of manganates are unstable, and are readily hydrolyzed to permanganates, with simultaneous precipitation of manganese dioxide. The change is brought about more readily by acids; even carbon dioxide may be used:

$$3K_2MnO_4 + 2CO_2 = 2KMnO_4 + 2K_2CO_3 + MnO_2.$$

Chlorine oxidizes manganates quantitatively to permanganates:

$$2K_2MnO_4 + Cl_2 = 2KMnO_4 + 2KCl.$$

Permanganates possess considerably greater stability than manganates. Certain oxidizing agents can effect the transformation of manganous salts directly into permanganates, and such reactions are sometimes used for identifying the element. If a ✳ drop of a solution of a manganous salt is stirred on a white tile with two drops of concentrated sulphuric acid and a drop of silver nitrate together with a little solid ammonium persulphate, the violet colour of the permanganate ion appears after a short time. The presence of a trace of the silver salt is necessary to catalyse the oxidizing action of the persulphate.

In presence of dilute sulphuric acid, a number of substances can be oxidized quantitatively by a solution of potassium permanganate. Such reactions are regularly employed for the purposes of volumetric analysis. Thus ferrous salts are oxidized to ferric salts, oxalic acid is oxidized to carbon dioxide and water, ferrocyanides to ferricyanides, vanadous salts or vanadyl salts to vanadates, nitrites to nitrates. In such reactions, the permanganate is exerting its maximum oxidizing action, being reduced to a manganous salt, two molecules of the salt giving rise to five atoms of available oxygen, thus:

$$2KMnO_4 + 3H_2SO_4 = K_2SO_4 + 2MnSO_4 + 3H_2O + 5O,$$

or as an ionic equation:

$$\overset{-}{Mn}O_4 + 8\overset{+}{H} + 5e = \overset{++}{Mn} + 4H_2O.$$

In neutral, or feebly acid solution, permanganates react with manganous salts according to the so-called Guyard reaction with precipitation of manganese dioxide, thus:

$$2KMnO_4 + 3MnSO_4 + 2H_2O = 5MnO_2 + K_2SO_4 + 2H_2SO_4.$$

Potassium permanganate is sometimes used as an oxidizing agent in alkaline (sodium carbonate) solution. Under these conditions, two molecules of the permanganate give rise to three atoms of available oxygen, reduction proceeding to the stage of manganese dioxide. Formic acid is oxidized quantitatively to carbon dioxide and water under these conditions.

ZINC

Ammonium sulphide precipitates white zinc sulphide, ZnS, insoluble in excess of the reagent and in very weak acids such as acetic acid. It is, however, readily soluble in strong acids. Hydrogen sulphide precipitates zinc sulphide completely from a solution of a zinc salt if sufficient sodium acetate is present to diminish the effective acidity of the solution.

Sodium hydroxide precipitates colourless zinc hydroxide, $Zn(OH)_2$, readily soluble in excess of the reagent, in acids, and in ammonia. The solubility in acids is due to the formation of zinc salts. The solubility in excess of sodium hydroxide is due to the formation of sodium zincate, $\overset{+}{Na}[\overset{-}{Zn}(OH)_3]$, in which the zinc is present as a complex anion; while the solubility in ammonia rests on the formation of the complex cation, $Zn(\overset{++}{NH_3})_4$.

Neither complex ion has great stability, and consequently hydrogen sulphide precipitates zinc sulphide completely from a solution of sodium zincate or from a solution of the ammine hydroxide.

Sodium carbonate precipitates basic zinc carbonate, insoluble in excess of the reagent. On ignition, the basic salt is converted into zinc oxide. Zinc is sometimes determined gravimetrically in this way.

✣   *Mercuric chloride* in presence of excess of *ammonium* or *potassium thiocyanate* precipitates white zinc mercuric thiocyanate, $Zn[Hg(SCN)_4]$.

Potassium ferrocyanide precipitates a sparingly soluble potassium zinc ferrocyanide, $K_2Zn_3[Fe(CN)_6]_2$. This reaction has been adapted for determining zinc by a volumetric method.

When zinc oxide is moistened with a solution of *cobalt nitrate* and heated strongly, a fine green infusible mass, known as Rinmann's green, is obtained. The product consists of mixed crystals of zinc oxide and cobaltous oxide. The test is best carried out by moistening a filter paper with a mixture of the solution to be examined and cobalt nitrate, and then calcining the paper. *Potassium cobalticyanide* may, with advantage, be used instead of cobalt nitrate for carrying out this test. According to Feigl (*Qualitative Analyse mit Hilfe von Tüpfelreaktionen,* 1935, p. 235) the best conditions for working consist in mixing a dilute nitric acid solution of the substance to be tested with a solution of potassium cobalticyanide. A filter paper is soaked in the solution, zinc cobalticyanide being precipitated on the paper. After rejecting the unused liquid, the paper is dried over a flame and ignited. In zinc cobalticyanide, the ratio of zinc to cobalt is the most favourable for the production of the green mixed crystals of zinc and cobaltous oxides. It is definitely advantageous to carry out the test in this way, because the carbon of the filter paper impedes the formation of cobaltic oxide, which has a strong red-brown colour and might mask the green crystals. Similar considerations apply to the corresponding test for magnesium (see p. 75).

VANADIUM

This element exists in four well-defined states of oxidation, corresponding to the oxides, V_2O_2, V_2O_3, V_2O_4 and V_2O_5. The ions derived from these oxides have distinctive colours. The bivalent vanadous ion is violet, the tervalent vanadic ion is green, the salts derived from the tetroxide, V_2O_4, contain the complex bivalent vanadyl ion $\overset{++}{V}O$, which has a fine blue colour, and lastly anions derived from the pentoxide, V_2O_5, of which the metavanadates are the simplest and contain the yellow anion $\overset{-}{V}O_3$. In addition to these, pervanadates are known.

The vanadous salts are characterized by very powerful reducing properties. Powerful *reducing agents*, such as *zinc* and *dilute sulphuric acid*, convert vanadates into vanadous salts, the successive colour changes from yellow through blue to green and finally to violet being well marked. Milder reducing agents, such as *ferrous sulphate, sulphur dioxide* or *hydrogen sulphide*, convert vanadates into the blue vanadyl salts, but further reduction does not take place. *Potassium permanganate* in presence of *dilute sulphuric acid* will oxidize vanadium in any of the lower stages to vanadic acid. *Caro's acid* will also oxidize vanadyl salts to vanadic acid at the ordinary temperature. *Hydrogen peroxide* will convert vanadates in acid solution into pervanadates which have a brown colour. These various colour changes, which are concerned with oxidation and reduction, are of value for recognizing the element, and some, such as the oxidation of vanadous salts or vanadyl salts to vanadic acid, are of importance in volumetric analysis. Vanadic anhydride is soluble both in acids and in alkalis, more easily in the latter. The colour in alkaline solution is pale yellow, in acid solution the colour is distinctly darker. Complex ion formation doubtless takes place.*

* The chemistry of vanadic acid is complicated. Vanadic anhydride, V_2O_5, obtained by igniting ammonium metavanadate, NH_4VO_3, is an orange-red powder, extremely sparingly soluble in water to a pale yellow solution with an acid reaction to indicators. It is more readily soluble in alkalis than in acids. Physico-chemical methods indicate that the solutions of vanadates rarely contain the simple $\overset{-}{V}O_3$ anion; more complex anions being present.

None of the ordinary group reagents produces precipitates with vanadium compounds. However, when *ammonium sulphide* is added to a solution of a vanadate, the liquid assumes a red colour, due to the formation of a thiovanadate. When such a solution is acidified, the brown pentasulphide, V_2S_5, is precipitated. At the same time, a certain amount of reduction takes place, the solution turning blue in consequence of the production of a vanadyl salt.

Although vanadium remains unprecipitated by the ordinary group reagents when a pure vanadium salt is analysed, such is not the case if other elements, such as aluminium, iron or chromium, are present. In such cases, a considerable amount of vanadium is precipitated together with these metals when they are precipitated as hydroxides by the action of ammonia. The presence of vanadium in the ammonia precipitate may be recognized by dissolving a portion of the precipitate in dilute sulphuric acid and applying special tests. In this connexion, it must be remembered that cerium and titanium, as well as some other elements, interfere with the colour produced by the action of hydrogen peroxide.

BARIUM

Ammonium carbonate precipitates white barium carbonate, $BaCO_3$. This compound is appreciably more soluble in water than strontium and calcium carbonates. It is sometimes used to separate the more weakly basic tervalent hydroxides, such as ferric hydroxide, from the relatively stronger bivalent ones, such as nickel hydroxide (see p. 125).

Hydrofluosilicic acid precipitates the colourless crystalline fluosilicate, $BaSiF_6$, especially if alcohol is added to the solution. This reaction is distinctive for barium, and is not given by strontium or calcium salts.

Dilute sulphuric acid produces, even from very dilute solutions, a white precipitate of barium sulphate, $BaSO_4$, insoluble in acids. The solubility of barium sulphate in water is about one part of the salt to 400,000 parts of water. On this account, it is used for the gravimetric determination of barium, and reciprocally, of sulphuric acid and sulphur generally.

Potassium chromate or *dichromate* precipitates yellow barium chromate, $BaCrO_4$, soluble in mineral acids, but insoluble in acetic acid.

Ammonium oxalate precipitates barium oxalate, which is similar in properties to the oxalates of calcium and strontium as regards its insolubility in ammoniacal liquids and its solubility in acids.

Barium chloride and nitrate are both insoluble in absolute alcohol. If concentrated hydrochloric acid be added to an aqueous solution of barium chloride, much of the salt is precipitated, but may be redissolved by diluting with water.

Barium compounds impart a characteristic apple-green colour to the non-luminous flame, which persists for a long time.

STRONTIUM

Ammonium carbonate precipitates white strontium carbonate, $SrCO_3$. Its properties are similar to those of calcium carbonate.

Ammonium oxalate in presence of ammonium chloride and ammonia precipitates the oxalate, SrC_2O_4, which resembles the corresponding calcium compound.

Dilute sulphuric acid precipitates strontium sulphate after a short time. The solubility of strontium sulphate in water has been stated as one part of the salt in 7000 or 8000 parts of water at the ordinary temperature. Calcium sulphate may, with advantage, be used instead of dilute sulphuric acid for this test.

Potassium chromate produces no precipitate from dilute solutions of strontium salts, but from concentrated solutions a yellow precipitate of strontium chromate, $SrCrO_4$, may separate. Strontium chromate is readily soluble in acids, including acetic acid.

Strontium chloride, nitrate and acetate resemble the corresponding calcium salts as regards their solubility in water. Strontium chloride is soluble in absolute alcohol, but the nitrate is practically insoluble.

Strontium compounds impart an intense crimson colour to the non-luminous flame.

CALCIUM

Ammonium carbonate precipitates calcium carbonate, $CaCO_3$, insoluble in excess of the reagent. In solutions containing ammonium chloride, the precipitation of calcium carbonate is very liable to be incomplete, owing to reversibility of the reaction, particularly if the solution is boiled. Indeed, if calcium carbonate is boiled with a solution of ammonium chloride, there is sufficient acid hydrolysis to result in the calcium going completely into solution as calcium chloride, ammonia and carbon dioxide being evolved.

Ammonium oxalate in presence of ammonium chloride and ammonia precipitates calcium oxalate quantitatively as a white micro-crystalline powder. This reaction is applied to the gravimetric determination of calcium, the oxalate being ignited and weighed as calcium carbonate, or more usually as calcium oxide.

Dilute sulphuric acid or a soluble sulphate produces no precipitate from *dilute* solutions of calcium salts. From *concentrated* solutions, however, calcium sulphate (gypsum), $CaSO_4, 2H_2O$, separates as a white precipitate. Gypsum is soluble in water to the extent of approximately one part in 500 parts of water at the ordinary temperature.

Most calcium salts are sparingly soluble or insoluble in water, but the halides and the nitrate are very soluble. Calcium chloride and calcium nitrate are both readily soluble in alcohol. The solubility of the calcium salts of some organic acids, e.g. calcium citrate, diminishes with rise of temperature.

Calcium salts impart a dull red colour to the non-luminous flame.

MAGNESIUM

Ammonia precipitates the hydroxide, $Mg(OH)_2$, and *ammonium carbonate* precipitates the basic carbonate, $MgCO_3xMg(OH)_2$, but neither of these reagents will give any precipitate in solutions containing ammonium chloride.

Hydrogen disodium phosphate, or better, *diammonium phosphate*, produces in solutions containing ammonium chloride and ammonia a white crystalline precipitate of magnesium ammonium phosphate, $MgNH_4PO_4$. This compound is very sparingly soluble in water, still less soluble in solutions containing ammonia, but readily soluble in acids. Magnesium is determined gravimetrically in this way, the precipitate being ignited and weighed as the pyrophosphate, $Mg_2P_2O_7$.

When magnesium oxide is ignited with *cobalt nitrate*, a faint but characteristic pink mass, consisting of mixed crystals of magnesium and cobaltous oxides, is obtained. As in the case of zinc, it is best to carry out the test by soaking a filter paper in a solution of a mixture of the substance to be tested with *cobalt nitrate*, or better, *potassium cobalticyanide*, and calcining the paper.

Of the various organic reagents which have been suggested ✷ for identifying magnesium, mention may be made of *paranitrobenzeneazoresorcinol*, which produces an intense blue adsorption product with magnesium hydroxide. The reaction may be carried out on a spot plate with an extremely dilute solution of the reagent in alcohol followed by a drop of dilute sodium hydroxide, or a very dilute aqueous solution containing alkali may be used. This procedure is preferable to using filter paper, as even magnesium-free filter paper produces a violet colour. Ammonia and ammonium salts must be absent.

POTASSIUM

Potassium salts impart a violet colour to the non-luminous flame. The colour is, however, masked by small traces of sodium. If, however, the flame be viewed through cobalt-blue glass, the violet-red flame of potassium is readily visible.

Tartaric acid produces, especially if crystallization is aided by scratching the vessel with a glass rod, a white crystalline precipitate of the bitartrate, $C_4H_4O_6KH$.

Picric acid precipitates the sparingly soluble yellow crystalline picrate, $C_6H_2(NO_2)_3OK$.

Perchloric acid precipitates white potassium perchlorate, $KClO_4$.

Sodium cobaltinitrite, $Na_3[Co(NO_2)_6]$, produces in neutral or
✿ weak acetic acid solutions a yellow precipitate of potassium
sodium cobaltinitrite. Under ordinary conditions, the precipitate has a composition approximating to the formula

$$K_2Na[Co(NO_2)_6],$$

but its composition is variable.

Chloroplatinic acid precipitates the yellow platinichloride,
K_2PtCl_6. This salt is employed for determining potassium gravimetrically. The element is also sometimes determined as the
perchlorate.

It should be noted that most of these reagents produce
sparingly soluble ammonium salts. Confusion may therefore
arise unless ammonium salts, if present, are removed by volatilization.

SODIUM

The characteristic yellow flame coloration of sodium is very
distinctive, but should not be relied upon unless other elements
are rigidly excluded, as very small quantities of the element in
the presence of large quantities of other substances produce a
strong coloration. Confirmation by a precipitation reaction is
therefore desirable.

Dihydroxytartaric acid, $\begin{matrix} C(OH)_2COOH \\ | \\ C(OH)_2COOH \end{matrix}$, produces a sparingly

soluble white precipitate of the disodium salt. Precipitation is
more complete if the dihydroxytartaric acid is exactly neutralized with potassium carbonate. As dihydroxytartaric acid is
oxidized quantitatively by potassium permanganate in acid
solution to carbon dioxide and water, a volumetric method for
determining sodium is available.

Acid potassium pyroantimonate, $K_2H_2Sb_2O_7$, produces a white
precipitate of the sparingly soluble sodium salt, $Na_2H_2Sb_2O_7$.
✿ *Zinc uranyl acetate* in dilute acetic acid solution produces a
yellow crystalline precipitate of sodium zinc uranyl acetate,

$NaZn(UO_2)_3(CH_3CO_2)_9$, $9H_2O$. The production of this precipitate is much aided by scratching the vessel with a glass rod.

LITHIUM

Lithium salts impart an intense carmine-red colour to the non-luminous flame.

Disodium hydrogen phosphate precipitates white lithium phosphate, Li_3PO_4, on boiling, but precipitation is incomplete unless a sufficient amount of alkali, best added as dilute sodium hydroxide, is present to neutralize the hydrogen ions which are set free in the reaction. Precipitation is impeded by the presence of ammonium salts, in consequence of the tendency to form complex ions.

Sodium carbonate precipitates lithium carbonate, Li_2CO_3, from moderately concentrated solutions, as a white precipitate, especially on boiling. The precipitate, like lithium phosphate, is soluble in solutions of ammonium salts, due to the formation of complex ions.

Sodium fluoride precipitates white lithium fluoride, LiF, at the ordinary temperature.

Dihydroxytartaric acid precipitates the sparingly soluble lithium dihydroxytartrate. This salt is somewhat more soluble in water than the corresponding sodium salt.

It will be noted that lithium is somewhat exceptional when compared with the other alkali metals. This behaviour is typical of the first member of a group in the periodic table. Lithium chloride, unlike the chlorides of the other alkali metals, is readily soluble in a number of organic solvents, such as alcohols, acetone, and pyridine. Advantage is taken of this in effecting the separation of lithium from these elements and in purifying the salts.

AMMONIUM SALTS

All ammonium salts when heated with alkalis are decomposed with evolution of ammonia, which is easily recognized by its pungent odour and alkaline properties.

Small quantities of ammonia or of ammonium salts are best

✵ recognized by the use of *Nessler's reagent*. This reagent is an aqueous solution of *potassium mercuric iodide*, prepared by adding potassium iodide to a solution of mercuric chloride until the precipitate of mercuric iodide is redissolved, containing excess of *potassium hydroxide*. In the presence of ammonia or of an ammonium salt, a brown coloration or precipitate is produced. The product is the iodide of the so-called Millon's base, and is formed according to the equation:

$$2\overline{Hg}\overline{I}_4 + NH_3 + 3\overline{OH} = [OHg_2NH_2]I + 7\overline{I} + 2H_2O.$$

This test is extremely sensitive and is frequently employed for the determination of ammonia colorimetrically.

Ammonium salts resemble potassium salts in some respects, as, for example, in giving sparingly soluble platinichlorides and bitartrates with solutions of chloroplatinic and tartaric acids respectively. It is therefore necessary to eliminate ammonium salts when applying these reagents as tests for potassium. All ammonium salts volatilize when heated, the halides suffering reversible dissociation into the halogen hydride and free ammonia, but the nitrate decomposes into nitrous oxide and water.

Chapter IV

REACTIONS OF THE ACID RADICALS

(Highly sensitive reactions suitable for spot tests
are indicated by asterisks ✿)

CARBONATES

Dilute hydrochloric acid decomposes carbonates with evolution
of carbon dioxide, which is usually recognized by its reaction
with lime water. A solution of barium hydroxide (baryta water)
is more sensitive. Traces of carbon dioxide are best detected in
the absence of acid gases with the aid of phenolphthalein. A
very dilute solution of sodium carbonate is used, of about $N/500$
or $N/1000$ concentration, containing a small quantity of phenolphthalein dissolved in alcohol. In presence of carbon dioxide,
sodium bicarbonate is formed, the pH value of which at 8·35
corresponds to the decolorization of the indicator.

The metallic carbonates, apart from those of the alkali metals,
are insoluble in water; the bicarbonates are soluble. Precipitates
are therefore produced by adding solutions of the heavy metals.
The normal carbonates are, however, rarely precipitated, basic
carbonates are more commonly produced. If *silver nitrate* be
added to a solution of an alkaline carbonate, silver carbonate,
Ag_2CO_3, separates as a white precipitate, but if the liquid is
heated, hydrolysis takes place and the dark brown oxide, Ag_2O,
separates. Similar phenomena are observed if copper sulphate
is added to a solution of a carbonate. At first the basic greenish
blue cupric carbonate is precipitated, but on heating the black
oxide is formed.

It should be remembered that acids liberate carbon dioxide
from cyanates as well as from carbonates. The cyanic acid which
is liberated is hydrolyzed with formation of an ammonium salt:

$$KCNO + 2HCl + H_2O = KCl + NH_4Cl + CO_2.$$

FLUORIDES

Heated with *concentrated sulphuric acid*, fluorides evolve hydrogen fluoride. The gas is characterized by its property of reacting with silica to form silicon tetrafluoride, SiF_4, and on this account it etches glass. For qualitative purposes, the test may be carried out by decomposing the fluoride in a test tube cleansed with a little warm concentrated sulphuric acid and a trace of powdered potassium dichromate just before adding the unknown substance. If hydrogen fluoride is not evolved, the acid mixture will flow smoothly down the tube, but if it is present drops will be left on the walls.

Most fluorides are insoluble in water, except the alkali fluorides and silver fluoride. The soluble salts are precipitated by solutions of many metallic salts, such as calcium chloride or barium chloride, but these reactions are not very characteristic.

Unlike the other halogen hydracids, hydrofluoric acid is a weak acid and the fluoride ion has a strong tendency to form complex ions. A characteristic test for fluorides depends upon the formation of a stable zirconium fluoride ion, $\overline{Zr}\overline{F}_6$, which is colourless.

✿ Solutions of *zirconium salts*, acidified with hydrochloric acid, produce an intense red-violet colour with a solution of *sodium alizarinsulphonate*. On adding a solution of a fluoride, the colour changes immediately to a pale yellow, due to the formation of the zirconifluoride ion. The test may be carried out conveniently on filter paper. The red colour of ferric thiocyanate is also bleached by fluorides, but as many other substances, e.g. mercuric chloride, behave similarly, the reaction is much less valuable for identifying fluorides. The reaction with zirconium alizarinsulphonate may even be used to characterize insoluble fluorides, such as calcium fluoride. The tendency to form the complex zirconifluoride ion is so strong that the acidified solution of the zirconium salt dissolves some of the calcium fluoride by withdrawal of fluoride ions and consequent transformation of the deep violet-red colour into a pale yellow.

The tendency of fluoride ions to form complex ions is evident in other directions. It has long been known that a fluoride with excess of a borate does not etch glass. This phenomenon is due to the formation of hydrofluoboric acid, $\overset{+}{H}[\overset{-}{BF_4}]$, the fluoborate anion having considerable stability. A similar derivative, hydrofluosilicic acid, $\overset{++}{H_2}[\overset{--}{SiF_6}]$, is obtained by distilling a mixture of finely powdered silica or glass, calcium fluoride, and concentrated sulphuric acid and passing the vapours into water.

The hydrofluosilicic acid is produced together with hydrated silica by the hydrolysis of silicon tetrafluoride:

$$3SiF_4 + 2H_2O = SiO_2 + 2H_2SiF_6.$$

Hydrofluoboric and hydrofluosilicic acids are both stronger than hydrofluoric acid. When a solution of hydrofluosilicic acid is neutralized by a solution of sodium hydroxide in presence of a suitable indicator, neutrality is indicated when two equivalents of alkali have been added. After a short time, however, the solution develops acid properties and four equivalents of alkali are required to produce a stable neutral solution. These phenomena are due to the following reactions:

$$H_2SiF_6 + 2NaOH = Na_2SiF_6 + 2H_2O,$$

followed by

$$Na_2SiF_6 + 4NaOH = 6NaF + H_4SiO_4.$$

CHLORIDES

Heated with *concentrated sulphuric acid*, or other suitable acid of low volatility, hydrogen chloride is evolved.

When mixed with powdered *potassium dichromate* and heated with *concentrated sulphuric acid*, chromyl chloride, CrO_2Cl_2, a dark red volatile liquid having an appearance similar to bromine is liberated. This compound is the acid chloride of chromic acid, and its formation may be expressed by the equation:

$$2CrO_3 + 4HCl = 2CrO_2Cl_2 + 2H_2O.$$

If the compound is separated and treated with water, it is immediately hydrolysed to hydrochloric and chromic or dichromic acids:

$$2CrO_2Cl_2 + 3H_2O = H_2Cr_2O_7 + 4HCl.$$

6

When mixed with other oxidizing agents and heated with sulphuric acid, chlorine is evolved.

Aqueous solutions of chlorides acidified with nitric acid give a white precipitate of silver chloride, on adding *silver nitrate.* See silver, p. 29. The other sparingly soluble chlorides are those of lead, mercury (ous) and thallium (ous).

It should be noted that mercuric chloride and silver chloride do not evolve hydrogen chloride when heated with concentrated sulphuric acid. The presence of chlorine in these compounds can, however, be shown by reducing them with zinc and dilute sulphuric acid, filtering, and then testing the resulting filtrate with nitric acid and silver nitrate.

BROMIDES

Heated with *concentrated sulphuric acid*, a mixture of hydrogen bromide and bromine is evolved. In this reaction, the hydrogen bromide functions as a reducing agent, some of the sulphuric acid being reduced to sulphur dioxide:

$$2HBr + H_2SO_4 = 2H_2O + SO_2 + Br_2.$$

This reaction is reversible. A solution of hydrobromic acid may be prepared by passing sulphur dioxide through saturated bromine water.

No compound corresponding to chromyl dichloride is obtained when a bromide mixed with potassium dichromate is heated with sulphuric acid. The hydrogen bromide is simply oxidized to bromine. This difference in behaviour is sometimes used for testing qualitatively for chlorides in presence of bromides or iodides (which behave similarly to bromides).

Solutions of bromides acidified with nitric acid give with *silver nitrate* a pale yellow precipitate of silver bromide. This compound is still less soluble in water than silver chloride, and moreover is more sparingly soluble in aqueous ammonia. Silver bromide is considerably more sensitive to light than silver chloride.

As bromine is less electronegative than chlorine, solutions of bromides when treated with *chlorine water* liberate bromine, the liquid assuming a pale yellow or brown colour according to the quantity present. If the liquid be shaken with an organic solvent, such as *chloroform*, in which bromine is more soluble, the colour will be concentrated in the chloroform layer.

A very sensitive test for traces of bromine *vapour* consists in ☿ the use of a filter paper impregnated with a dilute solution of *fluorescein*. A trace of bromine converts this into eosin (tetrabromofluorescein), easily recognizable by the red colour.

IODIDES

Heated with *concentrated sulphuric acid*, vapours of iodine together with reduction products of sulphuric acid are obtained. Hydriodic acid is a much stronger reducing agent than hydrobromic acid, so very little free hydrogen iodide is evolved. The reduction products of the sulphuric acid are sulphur dioxide, sulphur and hydrogen sulphide.

Silver nitrate added to solutions of iodides acidified with nitric acid gives a yellow precipitate of silver iodide. This compound is still more insoluble than the bromide, and is also practically insoluble in aqueous ammonia. It is less sensitive to light than either silver chloride or bromide.

Since iodine is the least electronegative of the halogens, iodine is displaced from aqueous solutions of iodides on adding either *chlorine water* or *bromine water*. If the liquid be shaken up with *chloroform*, the iodine, being more soluble in chloroform than in the aqueous solution, becomes largely concentrated in the chloroform layer with production of a violet colour. If chlorine water be gradually added in excess after all the iodine has been displaced, union of the two halogens takes place with production of iodine monochloride, the violet colour of the iodine in the chloroform vanishing. These reactions may be adapted to the detection of a bromide and an iodide together in *dilute* solution. A few drops of chlorine water are first added, and the liquid shaken with a little chloroform, which immediately turns violet. Chlorine water is then added, with thorough shaking between each addition, until the chloroform layer becomes colourless, indicating complete transformation of the iodine into iodine monochloride. On further addition of chlorine water, the chloroform layer assumes a brown colour due to liberation of bromine from the bromide. Very dilute solutions must be used for this test.

The fluorescein test for bromine vapour is inapplicable in the presence of iodine, since tetraiodofluorescein (erythrosin) which has a dark red-violet colour would be produced. Iodine vapour produces a deep blue colour with starch paste.

Silver iodide, bromide and chloride are all sufficiently insoluble in water for the purposes of gravimetric determination of these halogens. Nevertheless, there are very considerable differences in the solubilities. The values, expressed in moles per litre at 25° C., are 1.0×10^{-8} for the iodide, 0.725×10^{-6} for the bromide, and 1.4×10^{-5} for the chloride. It is possible to precipitate silver iodide quantitatively in a solution containing a chloride and an iodide without any simultaneous precipitation of silver chloride on account of the very much greater insolubility of the iodide.* A sharp separation between an iodide and a bromide, or between a bromide and a chloride, cannot be effected in this way, but it is sometimes useful to adopt the method of fractional precipitation with silver nitrate in order to effect a partial separation of one particular halogen, when there is difficulty of detecting it in the presence of a great excess of another.

Cupric sulphate added to a solution of an iodide produces a precipitate of cuprous iodide mixed with iodine. If a reducing agent, such as *sulphurous acid*, is present, cuprous iodide alone separates as a dirty white precipitate. This reaction is of importance, as chlorides and bromides do not react in this way, and it enables iodides to be removed from the other halides:

$$2KI + 2CuSO_4 + H_2SO_3 + H_2O = 2CuI + 2KHSO_4 + H_2SO_4.$$

* The determination of an iodide in the presence of a chloride may be effected volumetrically, using Rose Bengal as an adsorption indicator.

HYPOCHLORITES

Dilute acids decompose hypochlorites with production of a solution of hypochlorous acid, HOCl, which has an odour similar to but not identical with that of chlorine. If dilute hydrochloric acid is used, the two acids react together with evolution of chlorine:

$$HCl + HOCl = H_2O + Cl_2.$$

Solutions of hypochlorites are unstable and gradually decompose, even at the ordinary temperature, with formation of chloride and chlorate. As hypochlorous acid is extremely weak, solutions of hypochlorites exhibit alkaline hydrolysis. On account of their instability, hypochlorites possess strong oxidizing and bleaching properties.

A few drops of a solution of a *cobalt salt* added to a solution of a hypochlorite results in the production of a precipitate of the higher cobalt oxide, and a stream of oxygen is evolved.

Sodium hypochlorite as ordinarily prepared by the action of chlorine on a solution of cold sodium hydroxide always contains chloride as well as hydroxide, and consequently gives the reactions of those ions, in addition to such as are distinctive of hypochlorites. A solution of pure sodium hypochlorite can only be prepared directly from a solution of the acid and sodium hydroxide.

Bleaching powder, which is obtained by the action of chlorine upon dry slaked lime, exhibits the characteristic reactions of hypochlorites, but it cannot be regarded as a simple hypochlorite. The compound is sometimes termed calcium chlorohypochlorite, and has been given, though with scanty justification, the empirical formula $CaOCl_2$. It is imperfectly soluble in water, and its chemical behaviour may possibly be explained on the basis of the decomposition of the chlorohypochlorite ions into chloride and hypochlorite ions.

An important derivative of hypochlorous acid which is much used as an antiseptic is the sodium compound of paratoluene-sulphochloramide, known as chloramine-T. This compound has the formula $C_6H_4\!\!\left\langle\begin{array}{c}CH_3\\SO_2\bar{N}\!\!\left\langle\begin{array}{c}\overset{+}{Na}\\Cl\end{array}\right.\end{array}\right.$, and it crystallizes with three mole-

cules of water. It is readily soluble in water, and the resulting solution possesses the distinctive properties and reactions of hypochlorites. Thus it oxidizes a solution of sodium arsenite quantitatively to sodium arsenate, and liberates iodine quantitatively from acidified potassium iodide. The solution is more stable than that of sodium hypochlorite.

CHLORATES

When heated, chlorates decompose with evolution of oxygen, and leave ultimately a residue of the chloride. If the heating is discontinued before decomposition is complete, some perchlorate is usually produced. If heated with an oxidizable substance such as charcoal, violent deflagration takes place.

Heated with *concentrated sulphuric acid*, a mixture of chlorine and oxygen from the explosive chlorine dioxide is produced. The chloric acid which is first liberated by the sulphuric acid decomposes into perchloric acid, water and chlorine dioxide:

$$3HClO_3 = HClO_4 + H_2O + 2ClO_2.$$

As this test is dangerous, very small quantities of the substances should be used.

Heated with *concentrated hydrochloric acid*, a mixture of chlorine and chlorine peroxide is evolved.

As all chlorates are soluble in water, no precipitation reactions are available. Many *reducing agents*, such as sulphur dioxide or ferrous sulphate, however, convert chlorates into chlorides, which can then be recognized by adding nitric acid and silver nitrate.

It may be remarked that chlorates are much more stable than hypochlorites; and although they contain more oxygen, they are less powerful oxidizing agents on account of their greater stability. Perchlorates are more stable than chlorates, and are very difficult to reduce to chlorides. Potassium perchlorate is very sparingly soluble in water.

BROMATES

When solid bromates are heated, they are decomposed with evolution of oxygen and leave a residue of bromide. If a bromate is mixed with an oxidizable substance such as charcoal and heated, deflagration takes place.

When heated with *acids*, bromic acid, $HBrO_3$, is liberated, but it rapidly decomposes into bromine, water and oxygen.

Mixed with a *bromide*, *dilute acids* decompose both salts with evolution of bromine, mutual oxidation and reduction taking place between the hydrobromic and bromic acids:

$$KBrO_3 + 5KBr + 6HCl = 6KCl + 3H_2O + 3Br_2.$$

Silver nitrate added to a solution of a bromate produces a white precipitate of silver bromate, $AgBrO_3$. This compound is very sparingly soluble in dilute nitric, acid, but is readily soluble in ammonia. From the ammoniacal silver solution, reducing agents such as sulphur dioxide precipitate silver bromide as a very pale yellow solid.

Bromates are intermediate in properties between chlorates and iodates, but resemble chlorates more closely, particularly as regards instability.

IODATES

Solid iodates when heated are decomposed with evolution of oxygen and formation of iodides. If *dilute hydrochloric acid* be added to the residue, iodine will be liberated unless the decomposition of the iodate is complete. This reaction takes place in consequence of the interaction between iodic and hydriodic acids:

$$HIO_3 + 5HI = 3H_2O + 3I_2.$$

The reaction can, however, be made to take a completely different course if the iodate is in excess of the iodide, and concentrated hydrochloric acid is present. Under these conditions, the iodine is oxidized to colourless iodine monochloride, hydrolysis of which is prevented by the excess of hydrochloric acid:

$$KIO_3 + 2KI + 6HCl = 3KCl + 3ICl + 3H_2O.$$

Concentrated hydrochloric acid when heated with iodates decomposes them with evolution of a mixture of chlorine and iodine trichloride:

$$KIO_3 + 6HCl = KCl + ICl_3 + Cl_2 + 3H_2O.$$

Silver nitrate added to a solution of an iodate precipitates white silver iodate, $AgIO_3$, sparingly soluble in dilute nitric acid, but readily soluble in ammonia in consequence of the formation of complex ions. If a reducing agent, such as sulphur dioxide, is added to the ammoniacal solution, a pale yellow precipitate of silver iodide is produced.

Barium chloride precipitates white barium iodate, $Ba(IO_3)_2$, which is sparingly soluble in dilute nitric acid.

Although iodic acid is monobasic, salts such as potassium biiodate, $KH(IO_3)_2$, are known. Such salts are most simply regarded as additive compounds of the normal iodate and iodic acid.

NITRATES

When heated, nitrates produce various products according to the nature of the metallic constituent. The alkali nitrates evolve oxygen and leave a residue of the nitrite. The nitrates of the heavy metals evolve a mixture of nitrogen tetroxide and oxygen leaving a residue of the metallic oxide. Ammonium nitrate decomposes into nitrous oxide and steam.

When *concentrated sulphuric acid* is heated with a nitrate, vapours of nitric acid are evolved. If a metal such as copper is added, fumes of oxides of nitrogen are produced.

All normal metallic nitrates are soluble in water. The nitrates of a few organic bases are sparingly soluble; in particular the base *nitron*, $C_{20}H_{16}N_4$, forms an extremely sparingly soluble nitrate. A solution of nitron in dilute acetic acid is therefore used as a reagent for the detection and sometimes for the gravimetric determination of nitrates.

The familiar "brown ring" test for nitrates depends upon the reduction of nitric acid by *ferrous sulphate* to nitric oxide. The nitric oxide dissolves in the excess of ferrous sulphate, forming the dark compound nitrosoferrous sulphate, $[Fe(NO)]SO_4$. The test is usually carried out by adding a small quantity of the solution of the substance to be tested to a concentrated solution of ferrous sulphate, then carefully pouring in some concentrated sulphuric acid so as to form a heavy liquid layer at the bottom of the test tube. A brown ring forms at the junction of the liquids. If the tube be shaken, the nitrosoferrous sulphate is decomposed in consequence of the heat produced by the dilution of the sulphuric acid, nitric oxide escapes and the colour disappears. The procedure may be modified with advantage when very small quantities of the substance only are available for test. A small crystal of ferrous sulphate is placed in a small porcelain crucible or on a white tile and is moistened with a drop of the solution to be tested. A drop of concentrated sulphuric acid is then added to the crystal. In the presence of a nitrate a brown ring will appear round the crystal:

$$6FeSO_4 + 3H_2SO_4 + 2HNO_3 = 3Fe_2(SO_4)_3 + 4H_2O + 2NO.$$

Brucine in presence of *concentrated sulphuric acid* produces a fine red colour. This test is usually carried out so as to form a coloured ring in a manner identical with that in which the "brown ring" test is applied, except that a solution of brucine in water is substituted for the ferrous sulphate.

Nitrates can be reduced to various products according to the particular reducing agent which is selected and to the conditions under which the reaction takes place. The oxidizing action of nitric acid is greatly dependent upon its concentration. Highly dilute nitric acid has very little oxidizing action, so little indeed that hydrogen can be collected from magnesium placed in nitric acid of $N/25$ concentration. With more concentrated nitric acid, various oxides of nitrogen are evolved when it reacts with metals. Highly concentrated nitric acid is a very powerful oxidizing agent, particularly if it contains dissolved nitrogen tetroxide. Nitrates when treated with *zinc* and *dilute acetic acid* are partially reduced to nitrites. When treated with *aluminium* or better with *Devarda's alloy* (45 per cent aluminium, 5 per cent zinc and

50 per cent copper) and sodium hydroxide, they are completely reduced to ammonia. This reaction is sometimes employed for determining nitrates by volumetric analysis.

NITRITES

Dilute hydrochloric acid decomposes nitrites with evolution of oxides of nitrogen. Presumably the unstable nitrous acid decomposes partly as follows:

$$2HNO_2 = H_2O + NO + NO_2.$$

At the same time autoxidation takes place with the formation of a certain amount of nitric acid:

$$3HNO_2 = HNO_3 + 2NO + H_2O.$$

A blue colour is usually visible in the solution during the decomposition.

Most nitrites are readily soluble in water. The least soluble salt is silver nitrite, which is soluble to the extent of about one part in 300 parts of water.

Nitrous acid is a weak acid. Its salts therefore exhibit appreciable alkaline hydrolysis. It is an extremely reactive substance, and is both an oxidizing and a reducing agent. Physico-chemical methods have shown it to be a weaker oxidizing agent than nitric acid, but as nitrous acid is so much less stable than nitric acid, its oxidizing action is much more pronounced. Thus it readily liberates iodine from potassium iodide, but is oxidized to nitric acid by potassium permanganate.

✲ Of the numerous colour reactions which are used for the detection of traces of nitrous acid, mention may be made more particularly of those with *guaiacol* and with *α-naphthylamine* and *sulphanilic acid*. For the former, a saturated solution of beechwood creosote in water may be used. If a small quantity of aqueous creosote solution be added to a very dilute solution of a nitrite, and the solution acidified with dilute *sulphuric acid*, an intense yellowish brown colour is produced, due to the production of the para-nitroso compound. This reaction will detect one part of nitrous acid in one million of water. The reaction with sulphanilic acid and α-naphthylamine, which depends upon the formation of an azo dye, is still more sensitive and is carried out as follows: Separate dilute solutions of sulphanilic acid and α-naphthylamine in dilute (about 30 per cent) acetic acid are prepared. The solution of α-naphthylamine is best prepared by boiling the base with water, decanting the clear liquid from the dark-coloured residue, and adding some acetic acid to this. The concentration of these solutions should be of the order of 0·1 per cent. The reaction is carried out by placing a drop of the solution to be tested on a tile or filter paper, then adding a drop of the solution of sulphanilic acid followed by a drop of the solution of α-naphthylamine. In the presence of a nitrite a fine red colour is obtained. The colour is due to the coupling of the α-naphthylamine with diazotized sulphanilic acid, the diazotization having been effected by the nitrous acid thus:

$$SO_3HC_6H_4NH_2CH_3COOH + HNO_2 = SO_3HC_6H_4N_2CH_3CO_2 + 2H_2O,$$
$$SO_3HC_6H_4N_2CH_3CO_2 + H C_{10}H_6NH_2 = SO_3HC_6H_4N_2C_{10}H_6NH_2 + CH_3COOH.$$

According to Feigl (*Qualitative Analyse*, p. 324) the sensitiveness of this test is so great as to enable one part of nitrous acid in five million parts of water to be detected. It is therefore too sensitive for ordinary purposes.

Another highly sensitive reaction for nitrites is the yellow ✿ colour which is obtained with a solution of *metaphenylene diamine* in presence of dilute hydrochloric or sulphuric acid.

SULPHIDES

Sulphides are decomposed by *acids* with evolution of hydrogen sulphide, but individual sulphides differ considerably as regards their resistance to decomposition. Manganous sulphide is decomposed by dilute acetic acid. Ferrous sulphide is decomposed by dilute hydrochloric acid, while the sulphides of antimony and lead require heating with concentrated hydrochloric acid to effect decomposition.

Hydrogen sulphide is easily recognizable by its odour, or by its action on a filter paper moistened with a drop of a solution of *lead acetate*, which is immediately blackened.

✿ *Sodium nitroprusside* in presence of *dilute sodium hydroxide* or *ammonia* gives an intense violet coloration with hydrogen sulphide. It is said that the colour is produced by the nitroprusside, $Na_2[Fe(CN)_5NO]$, reacting with the alkaline hydrosulphide thus:

$$Na_2[Fe(CN)_5NO] + 2NaSH = Na_4[Fe(CN)_5NOS] + H_2S.$$

The colour is destroyed by acids and by excess of alkali. Solutions of alkaline sulphides produce the colour directly, but free hydrogen sulphide does not do so in the absence of alkali.

Silver is readily blackened by hydrogen sulphide or by alkaline sulphides owing to formation of the sulphide.

When polysulphides are decomposed by acids, hydrogen sulphide is evolved and sulphur is precipitated simultaneously.

Oxidizing agents decompose sulphides with separation of sulphur. Even sulphur dioxide can effect the oxidation of hydrogen sulphide, sulphur being liberated from both compounds. A small quantity of pentathionic acid, $H_2S_5O_6$, is also produced.

SULPHITES

Dilute hydrochloric acid decomposes sulphites with evolution of sulphur dioxide. This gas is readily recognizable by its pungent odour and by its reducing properties. Among the many tests which have been used for detecting small quantities of the gas by its reducing action, the removal of the colour from a very dilute solution of potassium permanganate may be mentioned as highly sensitive.

Free sulphurous acid does not readily decolorize certain dyestuffs of the triphenylmethane series, such as *malachite green*, but neutral solutions of the sulphites do so readily. In applying a test of this kind, the solution must be neutralized with sodium bicarbonate and added to the dilute solution of the dyestuff, which is at once reduced to the colourless leuco compound of malachite green.

Zinc and *dilute hydrochloric acid* reduce sulphites to hydrogen sulphide, which is easily recognized by its blackening action upon lead acetate.

Barium chloride precipitates white barium sulphite, $BaSO_3$, which is readily soluble in dilute hydrochloric acid with evolution of sulphur dioxide.

A solution of *iodine* oxidizes sulphites to sulphates. Other oxidizing agents have an identical action.

Silver nitrate precipitates white silver sulphite, Ag_2SO_3. The precipitate is soluble in excess of the alkaline sulphite on account of the formation of a complex anion, $Ag\overline{SO}_3$. A solution containing the complex salt, however, gradually decomposes with separation of silver.

Sulphurous acid can in certain circumstances act as an oxidizing agent. Thus it oxidizes stannous chloride to the stannic condition when warmed in aqueous solution, yellow stannic sulphide being precipitated.

SULPHATES

Barium chloride in presence of *dilute hydrochloric acid* precipitates white barium sulphate, $BaSO_4$, insoluble in acids. This precipitate is formed even in very dilute solutions. Too great a concentration of hydrochloric acid is to be avoided, since barium chloride is liable to separate from solution. Further, if metals which produce sparingly soluble chlorides are present, the precipitation of barium sulphate should be effected with *barium nitrate* in presence of dilute *nitric acid*.

Lead acetate or *lead nitrate* precipitates white lead sulphate, $PbSO_4$. This compound is very sparingly soluble in water, still less so in dilute alcohol. It is, however, soluble in solutions of ammonium acetate or tartrate in consequence of the formation of complex ions.

Strontium nitrate precipitates strontium sulphate after a short time. It is considerably more soluble than barium sulphate. Calcium sulphate is only precipitated from concentrated solutions of sulphates.

When solid sulphates are ignited with *carbon*, they are reduced to sulphides. The resulting sulphide can then be decomposed by an acid and the hydrogen sulphide recognized. This test is, however, not specific for sulphates. Practically any sulphur compound can be reduced to a sulphide in this way. A more efficient method for decomposing some sulphur compounds consists in heating them with sodium, and then testing a solution of the resulting alkaline sulphide by the nitroprusside reaction.

THIOSULPHATES

Dilute hydrochloric acid decomposes thiosulphates with evolution of sulphur dioxide and separation of yellow sulphur. Under certain conditions, the sulphur is obtained in colloidal solution. Thus if a solution of sodium thiosulphate is acidified with concentrated sulphuric acid, the sulphur which is liberated readily assumes the colloidal condition.

Lead acetate precipitates white lead thiosulphate, PbS_2O_3. When the solution is warmed, the precipitate blackens owing to the production of lead sulphide, which is formed by hydrolysis:

$$PbS_2O_3 + H_2O = PbS + H_2SO_4.$$

Silver nitrate precipitates white silver thiosulphate, $Ag_2S_2O_3$, which is readily hydrolysed to silver sulphide with consequent blackening of the precipitate. Excess of sodium thiosulphate dissolves silver thiosulphate in consequence of the formation of

stable complex anions, such as AgS_2O_3. The silver halides dissolve in sodium thiosulphate for the same reason.

Thiosulphates are strong reducing agents. The milder oxidizing agents, such as *iodine*, convert thiosulphates into tetrathionates, a reaction much used in volumetric analysis. *Ferric chloride* also oxidizes them to the tetrathionate stage with simultaneous reduction to a ferrous salt. During the reduction, an intense violet colour is produced due to the formation of a complex ferrithiosulphate ion. This reaction is catalysed by cupric salts (see p. 39). Strong oxidizing agents convert thiosulphates to sulphates. The action of *hydrogen peroxide* is of exceptional interest as this compound can oxidize thiosulphates to tetrathionates or to sulphates according to the experimental conditions. The oxidation to tetrathionates is catalysed by iodides, while that to sulphates is catalysed by ammonium molybdate. These reactions have been studied in detail by Abel, and have been found to proceed according to the equations:

$$2Na_2S_2O_3 + H_2O_2 + 2CH_3COOH = Na_2S_4O_6 + 2CH_3COONa + 2H_2O$$
and $$Na_2S_2O_3 + 4H_2O_2 = Na_2SO_4 + H_2SO_4 + 3H_2O.$$

It will be noticed that acetic acid is added in the first reaction. These reactions are instructive for demonstrating the specific nature of a catalyst in influencing a particular reaction. The course of the reactions may be shown by the aid of tests made with methyl orange and barium chloride. In the first reaction, the acidity of the solution diminishes as acetic acid is used up in the reaction, and practically no precipitate with barium chloride should appear, since barium tetrathionate is soluble in water. In the second reaction, the acidity increases, since free sulphuric acid is liberated, and an abundant precipitate is produced on adding barium chloride.

Zinc and *dilute hydrochloric acid* reduce thiosulphates with evolution of hydrogen sulphide.

Thiosulphates are decomposed on ignition. The products are usually a sulphate and a polysulphide. If the ignition is carried out with access to the air, some of the sulphur takes fire.

HYDROSULPHITES OR HYPOSULPHITES

These salts are derivatives of hydrosulphurous acid, $H_2S_2O_4$, which has not been isolated, and are obtained by the action of reducing agents, such as zinc, on the alkali bisulphites in presence of excess of sulphurous acid. Sodium hydrosulphite, $Na_2S_2O_4$, is an important reducing agent.

Solutions of the heavy metals, such as *silver nitrate, cupric sulphate* or *mercuric chloride* are at once reduced with separation of the metal. Sodium hydrosulphite is also much used as a reducing agent in organic work; for example dyestuffs such as *methylene blue* are at once reduced to the colourless leuco-compounds. A solution of sodium hydrosulphite containing excess of sodium hydroxide is also used as an absorbent for oxygen in gas analysis.

An important technical derivative of sodium hydrosulphite, known as rongalite, is obtained by the action of *formaldehyde* on the compound. Rongalite is the sodium salt of formaldehyde sulphoxylic acid, and its formation is expressed by the equation:

$$Na_2S_2O_4 + 2CH_2O + H_2O = CH_2(OH)SO_2Na + CH_2(OH)SO_3Na.$$

rongalite   formaldehyde sodium
bisulphite

Sulphoxylic acid, H_2SO_2, is unknown, but rongalite is much used as a technical reducing agent, and is preferable in some respects to sodium hydrosulphite on account of its greater stability.

There is no difficulty in distinguishing hydrosulphites from other salts of sulphur acids which have reducing properties, since the reducing action of the hydrosulphites is so much more powerful. A dilute solution of methylene blue is immediately bleached.

PERSULPHATES

These salts are derivatives of perdisulphuric acid, $H_2S_2O_8$, and have pronounced oxidizing properties. When heated in the dry state, they evolve oxygen and produce pyrosulphates:

$$2K_2S_2O_8 = 2K_2S_2O_7 + O_2.$$

Aqueous solutions undergo hydrolysis with loss of oxygen and production of free sulphuric acid:

$$2K_2S_2O_8 + 2H_2O = 4KHSO_4 + O_2.$$

A fresh solution of a persulphate produces no precipitate with barium chloride, since barium persulphate is soluble in water, but after a short time, or rapidly on heating, a precipitate of barium sulphate is formed from the free sulphate ion derived from the hydrolysis.

The oxidizing action of persulphates is frequently catalysed by silver salts, as, for instance, in the oxidation of chromic salts to dichromates. Potassium iodide liberates iodine slowly when added to a persulphate. They do not react with permanganates *directly*.

When persulphates are treated with concentrated sulphuric acid at 0° C. and the mass poured on to crushed ice, a solution of permonosulphuric acid, H_2SO_5 (Caro's acid), is produced. This compound has much more powerful oxidizing properties than perdisulphuric acid. Caro's acid liberates iodine *immediately* from potassium iodide, and bromine from potassium bromide. It also oxidizes vanadyl salts to vanadic acid at the ordinary temperature. The production of Caro's acid from perdisulphuric acid is due to hydrolysis:

$$H_2S_2O_8 + H_2O = H_2SO_5 + H_2SO_4.$$

After a time, further hydrolysis takes place with formation of sulphuric acid and hydrogen peroxide:

$$H_2SO_5 + H_2O = H_2SO_4 + H_2O_2,$$

which may be recognized by the test described on p. 114.

PHOSPHATES

With the exception of the alkali metal phosphates, all salts of this acid are insoluble or very sparingly soluble in water. Orthophosphates in which all of the available hydrogen of the acid has been replaced by metals are unaltered by heat, but if some unreplaced hydrogen is still present, pyrophosphates or metaphosphates are obtained. Similar results are obtained when metallic ammonium phosphates are ignited. Thus magnesium ammonium phosphate, $MgNH_4PO_4$, loses water and ammonia on ignition and leaves a residue of magnesium pyrophosphate, $Mg_2P_2O_7$. Microcosmic salt, sodium hydrogen ammonium phosphate, $NaNH_4HPO_4$, when ignited leaves a residue of sodium metaphosphate, $NaPO_3$.

As orthophosphoric acid is an acid of moderate strength, the reaction of solutions of the various sodium phosphates to indicators differs considerably. Dihydrogen sodium phosphate, NaH_2PO_4, is alkaline to methyl orange but acid to phenolphthalein, while hydrogen disodium phosphate is just alkaline to phenolphthalein. A mixture of these two salts in solution is frequently used as a buffer solution, having a pH value which may be adjusted between 4·5 and 9·0 according to the relative proportions of the two salts.

Insoluble phosphates are, generally speaking, readily soluble in dilute hydrochloric acid. On neutralization with ammonia, they are reprecipitated. Complications are therefore certain to arise in analysing a solution which contains a phosphate when examining the precipitate which is produced by ammonia for metallic hydroxides (see p. 123).

Magnesia mixture (a solution of magnesium sulphate, ammonium chloride, and ammonia) precipitates white magnesium ammonium phosphate, readily soluble in dilute hydrochloric acid and also in dilute acetic acid.

Silver nitrate produces a yellow precipitate of silver phosphate, Ag_3PO_4, readily soluble in ammonia and in dilute nitric acid. It is, however, insoluble or nearly so in acetic acid.

Ferric chloride precipitates yellowish white ferric phosphate, $FePO_4$, which is readily soluble in dilute hydrochloric acid, but is insoluble in acetic acid. The formation of ferric phosphate is of importance in effecting the separation of phosphoric acid.

Ammonium molybdate in large excess and in presence of nitric acid precipitates yellow ammonium phosphomolybdate,

$$(NH_4)_3PO_4, \ 12MoO_3, \ 6H_2O,$$

on heating. Phosphomolybdic acid is a fairly strong oxidizing agent, and is easily reduced to a lower state of oxidation which ☆ has an intense blue colour. A sensitive test for a phosphate, depending upon this property, may be carried out as follows: A drop of the solution to be tested is placed on filter paper together with a drop of a solution of *benzidine* in dilute acetic acid and a drop of a solution of *ammonium molybdate*. The filter paper is then exposed to the vapour of *ammonia*, when the development of a blue colour indicates the presence of a phosphate. The blue colour arises from two reactions, viz. the reduction of the ammonium phosphomolybdate which is formed to molybdenum blue, and also the blue colour of the oxidation product of benzidine.

Pyrophosphates and metaphosphates may be distinguished from orthophosphates by the production of *white* precipitates with *silver nitrate*, which have the formulae $Ag_4P_2O_7$ and $AgPO_3$ respectively. Metaphosphates moreover possess the property of coagulating *albumen*.

HYPOPHOSPHITES

The salts of hypophosphorous acid are characterized by powerful reducing properties. The acid has the molecular formula H_3PO_2, but it behaves solely as a monobasic acid. When solid hypophosphites are heated, spontaneously inflammable phosphine is evolved and a residue of pyrophosphate or metaphosphate remains. Free hypophosphorous acid is also decomposed by heat into phosphine and phosphoric acid:

$$2H_3PO_2 = PH_3 + H_3PO_4.$$

Copper sulphate in presence of free acid produces a spongy brown precipitate, which is said to consist of cuprous hydride. On warming the solution, hydrogen is evolved and copper separates.

Solutions of many of the heavy metals are reduced to the metallic condition by hypophosphites.

Potassium iodate in presence of dilute acid is reduced to iodide. The hydriodic and iodic acids subsequently react together with liberation of iodine:

$$HIO_3 + 3H_3PO_2 = 3H_3PO_3 + HI,$$
$$HIO_3 + 5HI = 3H_2O + 3I_2.$$

The sensitiveness of this reaction can, of course, be increased by using starch. Further, the reaction may be applied in the reciprocal way as a test for iodates.

BORATES

The majority of metallic borates are derivatives of pyroboric acid, $H_2B_4O_7$, or of metaboric acid, HBO_2. Salts of orthoboric acid are rarely if ever obtained, but esters, such as ethyl borate, $B(OC_2H_5)_3$, are easily prepared, and on account of their volatility are of importance in analysis for removing boric acid from mixtures. The best known salt is borax, $Na_2B_4O_7$, which normally crystallizes with ten molecules of water, and is fairly soluble in water. The aqueous solution shows strong alkaline hydrolysis. Borates other than those of the alkali metals are very sparingly soluble in water.

Boric acid is a crystalline solid, moderately soluble in water, and an extremely weak acid. So feeble are the acid properties of the compound that it turns litmus a wine red colour. If, however, a solution of a substance rich in hydroxyl groups such as mannite be added to boric acid, a complex mannitoboric acid of moderate strength is produced. When borax is heated, the water of hydration is expelled, considerable swelling taking place during the process. Ultimately a clear glassy mass of sodium metaborate is left.

Heated with *methyl* or *ethyl alcohol* and *concentrated sulphuric acid*, the corresponding methyl or ethyl esters are produced. These compounds burn with a green-edged flame. This test is best carried out in a small evaporating dish.

A solution of a borate, acidified with dilute hydrochloric acid, turns *turmeric paper* a reddish brown colour. When the paper is dried and moistened with dilute sodium hydroxide, the brown colour is replaced by a dark bluish black colour.

Silver nitrate precipitates white silver metaborate, $AgBO_2$, from a solution of borax in the cold. If the solution is heated, hydrolysis results in the separation of brown silver oxide. If very dilute solutions are used, the hydrolysis of the borax is sufficient to result in the precipitation of silver oxide at the ordinary temperature.

The various boric acids and their salts, unlike the various phosphoric acids, cannot be distinguished by qualitative tests. It would appear that the various acids are in mobile equilibrium in solution, the reactions

$$H_3BO_3 \rightleftharpoons HBO_2 + H_2O, \quad \text{and} \quad 4HBO_2 \rightleftharpoons H_2B_4O_7 + H_2O,$$

and possibly $4H_3BO_3 \rightleftharpoons H_2B_4O_7 + 5H_2O$, proceeding simultaneously.

As might be expected, boric acid has a marked tendency to form complex acids. Thus with hydrofluoric acid it forms hydrofluoboric acid, HBF$_4$, which is stronger than either of its constituent acids. So called heteropolyacids, similar to phosphomolybdic acid, are also formed with molybdic and tungstic acids.

SILICATES

The only soluble silicates are those of the alkali metals. These compounds are decomposed by acids with the separation of silica. The silicic acid sometimes remains in solution in the colloidal condition, but gelatinization ultimately takes place.

Insoluble silicates are not attacked by any acids except hydrofluoric acid. The great majority of the silicates belong to this class, and the solvent action of hydrofluoric acid upon such substances depends upon the formation of silicon tetrafluoride.

All insoluble silicates are decomposed by fusion with sodium carbonate, carbon dioxide being evolved and a sodium silicate formed. Thus when a felspar is fused with sodium carbonate, the following reaction takes place:

$$KAlSi_3O_8 + 3Na_2CO_3 = 3Na_2SiO_3 + KAlO_2 + 3CO_2.$$

In carrying out a decomposition of this kind, the mineral must be ground to an impalpable powder, and intimately mixed with six times its weight of sodium carbonate. The mixture is then heated in a platinum crucible over a blast lamp until evolution of carbon dioxide has ceased completely. After the mass has cooled, the contents of the crucible are extracted with concentrated hydrochloric acid, which results in the separation of the silicic acid, the other constituents of the mineral being in solution as chlorides.

When silicic acid is heated above 100° C. the water is removed and anhydrous silica is left. The product dissolves in hydrofluoric acid, and should leave no residue on evaporation:

$$SiO_2 + 4HF = SiF_4 + 2H_2O.$$

CARBIDES

These compounds are obtained by the action of carbon upon a metal at very high temperatures. The classification of carbides presents some difficulties, but the most practical way of so doing is to divide them into two groups according as to whether they are capable of decomposition by water or dilute acids or resist decomposition.

Carbides of the alkali and alkaline earth metals are rapidly hydrolysed by water at the ordinary temperature with evolution of acetylene:

$$CaC_2 + 2H_2O = Ca(OH)_2 + C_2H_2.$$

Aluminium carbide is very slowly hydrolysed by cold water, but fairly rapidly on heating with formation of aluminium hydroxide and methane:

$$Al_4C_3 + 12H_2O = 4Al(OH)_3 + 3CH_4.$$

Beryllium carbide behaves similarly to aluminium carbide.

Apart from the examples just quoted, there appears to be no simple rule regarding the nature of the gaseous products obtainable when a carbide is decomposed by an acid. Some carbides, such as the important compound cementite, Fe_3C, are remarkably stable towards acids.

CYANIDES

Dilute hydrochloric acid, or indeed practically any acid, decomposes all cyanides, except mercuric and silver cyanides, with evolution of hydrocyanic acid. The odour is fairly characteristic, though some persons apparently have difficulty in recognizing it. In any case, care should be taken to avoid inhaling it on account of its poisonous nature. A sensitive test for recognizing
✿ hydrocyanic acid consists in converting it into ammonium thiocyanate by allowing the vapour to come in contact with a little *yellow ammonium sulphide* on a filter paper. The paper may be conveniently placed over a dish in which the substance is being treated with dilute hydrochloric acid. On adding a drop of

ferric chloride and a drop of *hydrochloric acid* to the filter paper, the characteristic red coloration due to ferric thiocyanate is produced.

Silver nitrate when added in sufficient excess precipitates white silver cyanide, AgCN, insoluble in dilute nitric acid. At first a transient precipitate is produced on adding silver nitrate to a solution of a cyanide, but this readily dissolves in the excess of cyanide on shaking, on account of the formation of the very stable argenticyanide ion $\overline{Ag(CN)_2}$. On adding more silver nitrate, the precipitation of silver cyanide can be made complete. When ignited, silver cyanide is decomposed with evolution of cyanogen and leaves a residue of silver. The compound is, therefore, readily distinguished from the silver halides.

Cyanides are converted into ferrocyanides on adding *ferrous sulphate* and warming. When *ferric chloride* and *hydrochloric acid* are added to the resulting solution, ferric ferrocyanide (Prussian blue), $Fe_4[Fe(CN)_6]_3$, is produced.

Cyanides in solution may also be recognized by their rapid solvent action on *cupric sulphide*, resulting in the formation of colourless potassium cupricyanide, $K_2[Cu(CN)_4]$:

$$CuS + 4KCN = K_2[Cu(CN)_4] + K_2S.$$

This test may readily be carried out on filter paper. A dilute ✿ *cupric sulphate* solution, or better, one of *cupric ammine sulphate*, is dropped on the paper. The drops are blackened by exposure to hydrogen sulphide. On adding a few drops of a cyanide solution, the black spots are bleached.

Mercuric cyanide may be identified by heating the compound in a small tube and observing the flame of the combustible gas which is evolved. The colour of the cyanogen flame is pinkish violet, similar to that of peach blossom. It is better, however, to devise a means of detaching the cyanide constituent in an ionizable form. This may be done by adding excess of a solution of *potassium iodide*, which forms the complex salt potassium mercuric iodide, $K_2[\overset{++}{Hg}\bar{I}_4]$, and potassium cyanide:

$$Hg(CN)_2 + 4KI = K_2[HgI_4] + 2KCN.$$

The cyanide liberated in this way may be identified conveniently by decomposition with dilute hydrochloric acid, followed by fixation of the volatile hydrocyanic acid with yellow ammonium sulphide as thiocyanate, which gives the red colour with ferric chloride and hydrochloric acid as previously described.

Fused potassium cyanide is a useful reducing agent, and is frequently used for this purpose in metallurgical operations. It removes oxygen from metallic oxides on heating with formation of potassium cyanate and the metal.

CYANATES

Dilute hydrochloric acid decomposes cyanates with evolution of carbon dioxide and formation of a solution of ammonium chloride. The reaction involves hydrolysis:

$$KCNO + 2HCl + H_2O = KCl + CO_2 + NH_4Cl.$$

As carbon dioxide is readily evolved when cyanates are treated with dilute acids, these salts are liable to be confused with carbonates. They may, however, be distinguished from carbonates by adding a slight excess of *sodium hydroxide* to the residual solution after reaction with an acid, and boiling, when ammonia will be evolved from the ammonium salt which is one of the products of hydrolysis.

Silver nitrate precipitates white silver cyanate from solutions of cyanates, readily soluble in ammonia. It is also soluble with decomposition in acids.

Since the reaction between ammonium cyanate and urea is reversible, a solution of urea when boiled with a solution of silver nitrate gradually produces a precipitate of silver cyanate.

Cobalt nitrate produces a blue colour, due to the formation of a complex salt containing the anion $Co(\overline{CNO})_4$. A similar reaction is given with thiocyanates.

THIOCYANATES

Silver nitrate precipitates the white thiocyanate, AgSCN, insoluble in dilute nitric acid. This is used for determining silver in acid solution in volumetric analysis.

Copper sulphate in presence of a *reducing agent*, such as *sulphur* ☼ *dioxide* precipitates white cuprous thiocyanate, CuSCN. This compound is sometimes used for the gravimetric determination of copper.

Ferric chloride produces a deep blood-red coloration, which is supposed to be due to the formation of unionized ferric thiocyanate, $Fe(SCN)_3$. If a substance is present which is capable of withdrawing either ferric or thiocyanate ions the colour is bleached. Thus fluorides and oxalates which form complex ferrifluoride and ferrioxalate ions remove the red colour. A similar bleaching action is exerted by mercuric chloride in consequence of formation of the stable mercurithiocyanate ion. A solution of *ammonium mercurithiocyanate* produces sparingly soluble precipitates with cobalt and with zinc salts.

Powerful oxidizing agents, such as acidified potassium permanganate or potassium iodate in presence of concentrated hydrochloric acid, oxidize thiocyanates. Under ordinary conditions, the oxidation products are hydrocyanic and sulphuric acids:

$$HSCN + 3O + H_2O = HCN + H_2SO_4.$$

Cobalt nitrate in presence of *acetone* produces a fine blue colour.

Moderately concentrated sulphuric acid decomposes thiocyanates with evolution of carbonyl sulphide, a colourless, odourless inflammable gas:

$$KSCN + 2H_2SO_4 + H_2O = COS + (NH_4)HSO_4 + KHSO_4.$$

FERROCYANIDES

Heated with *concentrated sulphuric acid*, carbon monoxide is evolved:

$$K_4Fe(CN)_6 + 6H_2SO_4 + 6H_2O$$
$$= 6CO + 2K_2SO_4 + 3(NH_4)_2SO_4 + FeSO_4.$$

Heated with *dilute sulphuric acid*, hydrogen cyanide is evolved:

$$2K_4Fe(CN)_6 + 3H_2SO_4 = K_2Fe[Fe(CN)_6] + 3K_2SO_4 + 6HCN.$$

The ferrocyanides of the heavy metals are insoluble in water; several have characteristic colours.

Silver nitrate precipitates white silver ferrocyanide,

$$Ag_4Fe(CN)_6,$$

which is insoluble in dilute nitric acid and in ammonia.

Ferrous sulphate produces a pale blue precipitate of potassium ferrous ferrocyanide, $K_2Fe[Fe(CN)_6]$, which rapidly darkens owing to absorption of oxygen with formation of Prussian blue.

✿   *Ferric chloride* gives an intense blue precipitate of ferric ferro-cyanide (Prussian blue), usually formulated as $Fe_4[Fe(CN)_6]_3$, but it always contains potassium. A so-called soluble Prussian blue, which is really a colloidal solution of potassium ferric ferrocyanide, $KFe[Fe(CN)_6]$, may be obtained by dropping ferric chloride into excess of a solution of potassium ferrocyanide:

$$K_4Fe(CN)_6 + FeCl_3 = KFe[Fe(CN)_6] + 3KCl.$$

✿   *Copper sulphate* gives a chocolate coloured precipitate of cupric ferrocyanide, $Cu_2Fe(CN)_6$.

✿   *Uranyl acetate* precipitates dark brown uranyl ferrocyanide

$$(UO_2)_2[Fe(CN)_6].$$

Oxidizing agents, such as *chlorine* or *bromine*, oxidize ferro-cyanides to ferricyanides. With potassium permanganate in presence of dilute sulphuric acid, the reaction is quantitative and is used in volumetric analysis:

$$KMnO_4 + 5K_4Fe(CN)_6 + 4H_2SO_4$$
$$= 5K_3Fe(CN)_6 + 3K_2SO_4 + MnSO_4 + 4H_2O.$$

FERRICYANIDES

Heated with *concentrated sulphuric acid*, carbon monoxide and a little carbon dioxide are evolved:

$$2K_3Fe(CN)_6 + 11H_2SO_4 + 13H_2O$$
$$= 3K_2SO_4 + 6(NH_4)_2SO_4 + 2FeSO_4 + 11CO + CO_2.$$

Heated with *dilute sulphuric acid*, hydrogen cyanide is evolved:

$$2K_3Fe(CN)_6 + 6H_2SO_4 = 3K_2SO_4 + Fe_2(SO_4)_3 + 12HCN.$$

A fairly close similarity between the action of concentrated and dilute sulphuric acid upon ferrocyanides and ferricyanides will be observed. In both cases, dilute sulphuric acid results in the production of hydrogen cyanide. With the concentrated acid, however, ferrocyanides evolve carbon monoxide only, whereas ferricyanides produce carbon dioxide as well. Further, it will be noted that the iron in ferricyanides is reduced to ferrous sulphate when the concentrated acid is used, but not with the dilute acid. From the investigations of Adie and Browning and later of Bassett and Corbet, it would appear that the primary action of sulphuric acid upon both salts is to produce hydrogen cyanide, which escapes unchanged when dilute acid is used. When, however, the concentrated acid is used, the hydrocyanic acid is hydrolysed to formic acid, which is at once decomposed by the excess of sulphuric acid with consequent liberation of carbon monoxide. The production of carbon dioxide in the case of ferricyanides is due to the oxidation of some of the formic acid by the iron, present in the salt in the ferric condition.

Silver nitrate precipitates orange-coloured silver ferricyanide, $Ag_3Fe(CN)_6$, insoluble in dilute nitric acid, but soluble in ammonia.

Ferrous sulphate produces a deep blue precipitate of ferrous ✿ ferricyanide, $Fe_3[Fe(CN)_6]_2$ (Turnbull's blue), insoluble in acids.

Ferric sulphate or *chloride* produces no precipitate, but the colour of the solution becomes dark brown. Presumably the solution contains ferric ferricyanide. A dilute solution of *pure* potassium ferricyanide and a *pure* ferric salt is a sensitive reagent for reducing substances. A blue or green colour is at once produced, in consequence of the formation of either ferrous ferricyanide or of ferric ferrocyanide. It may be remarked that the relationships between Prussian blue and Turnbull's blue and the soluble modifications of these compounds are extremely complicated. Apart from the composition of the products being subject to variation with alterations of the conditions of formation, the possibilities of isomerism are considerable.

Ferricyanides have well-defined oxidizing properties. Thus

potassium iodide in presence of an acid is oxidized to iodine with simultaneous reduction of the ferricyanide to ferrocyanide. As potassium ferricyanide is easily obtainable in highly pure condition, has a high equivalent weight, and is very soluble in water, the compound has been used for standardizing solutions of sodium thiosulphate for volumetric analysis. Alkaline potassium ferricyanide is a powerful oxidizing agent, converting solutions of lead salts into lead dioxide, thallous salts into thallic oxide. When mixed with a solution of hydrogen peroxide, oxygen is rapidly evolved.

NITROPRUSSIDES

These salts are obtained by oxidizing ferrocyanides with nitric acid. The best known is sodium nitroprusside, $Na_2[Fe(CN)_5NO]$, which crystallizes with two molecules of water. It is a dark red solid, readily soluble in water, but the solution gradually decomposes, particularly on exposure to light.

Silver nitrate produces a pinkish precipitate of silver nitroprusside, $Ag_2[Fe(CN)_5NO]$, insoluble in dilute nitric acid, but soluble in ammonia.

✿   *Alkaline sulphides* produce an intense violet colour, probably due to the formation of a complex salt, $Na_4[Fe(CN)_5NOS]$, which is destroyed by acids (see sulphides, p. 92).

Acetone, in the presence of *dilute sodium hydroxide*, produces a red colour. On adding excess of glacial *acetic acid*, the colour deepens to a dark crimson. This reaction, which is not distinctive for acetone, but is given by many other ketones and aldehydes, or, more generally, by substances which contain a methylene group with a mobile hydrogen atom, is stated by Cambi to proceed as follows:

$$Na_2[Fe(CN)_5NO] + CH_3COCH_3 + 2NaOH$$
$$= Na_4[Fe(CN)_5NO:CHCOCH_3] + 2H_2O.$$

Ferrous sulphate produces a pink precipitate, presumably of ferrous nitroprusside, $Fe[Fe(CN)_5NO]$, but ferric chloride produces no precipitate. In this respect, nitroprussides resemble ferricyanides.

FORMATES

Heated with concentrated *sulphuric acid,* carbon monoxide is evolved without any charring.

Heated with concentrated *sulphuric acid* and a little *ethyl alcohol,* vapours of ethyl formate having an agreeable odour are evolved. The odour is somewhat different from that of ethyl acetate.

Ferric chloride added to a neutral solution of a formate produces a dark red colour due to the formation of a solution of ferric formate. On boiling the solution, hydrolysis takes place, some of the formic acid escaping in the steam and a precipitate of basic ferric formate separates.

Formates reduce solutions of certain heavy metals on heating. Alkaline solutions of *copper* are reduced to cuprous oxide, *ammoniacal silver nitrate* to the metal, and *mercuric chloride* is reduced first to mercurous chloride and afterwards to mercury.

Potassium permanganate in alkaline (sodium carbonate) solution is reduced on heating with separation of manganese dioxide, the reaction being quantitative, two molecules of the permanganate oxidizing three molecules of formic acid to carbon dioxide and water:

$$2KMnO_4 + 3CH_2O_2 = 2KHCO_3 + 2MnO_2 + CO_2 + 2H_2O.$$

Formates of the alkali metals when heated rapidly evolve hydrogen and leave a residue of oxalate. The alkaline earth formates evolve carbon monoxide as well as hydrogen and leave a residue of carbonate.

All normal formates are soluble in water, the cupric salt and the lead salt crystallize easily. Formic acid is a distinctly stronger acid than acetic acid.

ACETATES

The effect of heat upon acetates depends upon the nature of the metallic constituent. The alkaline earth acetates, such as calcium acetate, evolve acetone and leave a residue of calcium carbonate. A mixture of sodium acetate and soda lime evolves methane with production of sodium carbonate. Ammonium acetate heated by itself is largely decomposed into acetamide and water.

Heated with *concentrated sulphuric acid*, vapours of acetic acid are evolved which are easily recognized by the characteristic odour. Heated with concentrated sulphuric acid and a *small* quantity of *ethyl alcohol*, ethyl acetate, $CH_3COOC_2H_5$, is evolved, which has a fragrant odour.

A mixture of an alkali acetate and *arsenious oxide* when heated produces the nauseating odour of cacodyl oxide, $[(CH_3)_2As]_2O$. The odour is quite unmistakable, and its production serves as a sensitive test for an acetate, or reciprocally for arsenic. On account of the extremely poisonous nature of the cacodyl derivatives, the greatest care should be taken in carrying out this test.

Neutral solutions of acetates give a dark red colour with *ferric chloride*, due to the formation of ferric acetate. When the liquid is boiled, the iron is precipitated as basic ferric acetate, due to hydrolysis of the normal salt, some acetic acid escaping in the steam.

Silver nitrate produces, in fairly concentrated neutral solutions, a white crystalline precipitate of silver acetate, CH_3COOAg. The precipitate is soluble in hot water.

Oxalates

Concentrated sulphuric acid decomposes oxalates on gentle heating with evolution of a mixture of carbon monoxide and carbon dioxide. Charring does not occur.

When solid oxalates are heated, decomposition takes place with evolution of carbon monoxide and formation of carbonates and ultimately of oxides. The behaviour of free oxalic acid is different. On heating, the acid first loses its water of crystallization, then some carbon dioxide is lost with production of formic acid, but the reaction is by no means quantitative; some of the oxalic acid volatilizes unchanged.

Silver nitrate precipitates white silver oxalate, $Ag_2C_2O_4$, readily soluble in strong acids, but insoluble in acetic acid.

Silver oxalate decomposes explosively when heated, evolving carbon dioxide and leaving a residue of silver.

Calcium chloride precipitates white calcium oxalate, CaC_2O_4, insoluble in ammonia and in solutions of ammonium salts. Calcium oxalate is soluble in mineral acids, but insoluble in acetic acid.

Barium chloride precipitates barium oxalate, which has properties similar to those of calcium oxalate.

Apart from the oxalates of the alkali metals, oxalates are insoluble in water. Most oxalates are, however, readily soluble in dilute solutions of strong acids. An important exception to this, however, is shown by the oxalates of the rare earths. The rare earth oxalates are precipitated on adding oxalic acid to a hydrochloric acid solution containing these metals; indeed oxalic acid might be described as the group reagent for the rare earths. The oxalates of zirconium, hafnium, and thorium differ from those of the rare earths in being soluble in excess of ammonium oxalate, due to the formation of complex oxalato salts.

Oxalic acid is somewhat stronger than most other carboxylic acids, and is dibasic. It has, however, the property of forming acid salts, in which it would at first sight appear to be tetrabasic. The well-known salt potassium quadroxalate,

$$KHC_2O_4, \ H_2C_2O_4, \ 2H_2O,$$

may be regarded as an addition compound of the binoxalate and free oxalic acid. Its tendency to form complex ions is well marked. In potassium ferrioxalate, $K_3[Fe(C_2O_4)_3]3H_2O$, the complex ferrioxalate anion, $\overline{Fe(\overline{C_2O_4})_3}$, is sufficiently stable to produce very little colour with the thiocyanate ion; it is, however, decomposed by ammonia with separation of ferric hydroxide. The chromioxalate anion, $\overline{Cr(\overline{C_2O_4})_3}$, is still more stable; at the ordinary temperature, a solution of potassium chromioxalate gives neither the reactions of a chromic salt nor of an oxalate.

Powerful oxidizing agents are reduced by oxalic acid. Thus with acidified potassium permanganate, quantitative reduction takes place, the oxalic acid being oxidized to carbon dioxide and water. Ceric salts are reduced to cerous salts, and if the oxalic acid is present in excess, cerous oxalate separates as a white precipitate.

TARTRATES

When solid tartrates are heated, charring takes place with evolution of volatile products which have an odour recalling that of burnt sugar. The alkali metal tartrates leave a residue of carbonates. When tartaric acid is heated with *potassium bisulphate*, pyruvic acid, $CH_3COCOOH$, together with water and carbon dioxide are produced.

Heated with *concentrated sulphuric acid*, a mixture of carbon monoxide, carbon dioxide, and sulphur dioxide is evolved. Charring rapidly takes place.

Potassium acetate in presence of *acetic acid* produces in solutions of tartrates a white crystalline precipitate of potassium bitartrate, $\begin{array}{l} CH(OH)COOK \\ | \\ CH(OH)COOH \end{array}$, readily soluble in acids. Crystallization is aided by scratching the vessel with a glass rod.

Calcium chloride produces in neutral solutions of tartrates a white precipitate of calcium tartrate, $CaC_4H_4O_6$, readily soluble in acids.

Ammoniacal silver nitrate, prepared by adding sodium hydroxide to a solution of silver nitrate, followed by dilute ammonia until the silver oxide is *nearly* dissolved, produces a deposit of silver in mirror condition on the walls of the test tube when warmed with a solution of a tartrate. The tartrate is functioning simply as a reducing agent, so this test is in no sense a specific one.

If a drop of *ferrous sulphate* followed by *hydrogen peroxide* and finally by excess of *sodium hydroxide* is added to a solution of tartaric acid or to that of a tartrate which does not contain heavy metals, an intense violet colour is produced. The colour is produced in consequence of the oxidation of the tartaric acid to dihydroxymaleic acid, $\begin{array}{l} C(OH)COOH \\ \| \\ C(OH)COOH \end{array}$, by the action of the hydrogen peroxide in the presence of the trace of ferrous salt, and the subsequent interaction between the iron (now in the ferric condition) with the dihydroxymaleic acid.

Tartrates have a very marked tendency to form complex ions with certain metallic ions, in many of which the distinctive reactions of the metals are partially obscured. Thus antimonious oxide is readily soluble in a solution of potassium bitartrate with formation of potassium antimonyl tartrate, a salt which is much less hydrolysed than the inorganic salts of antimony. The well-known Fehling's solution consists of a mixture of copper sulphate, Rochelle salt (potassium sodium tartrate), and potassium hydroxide. Cupric hydroxide is not precipitated, because the copper is present in solution as the dark blue cupritartrate ion. If ammonia be added to a solution of a ferric salt which contains tartrate ions ferric hydroxide is not precipitated, because the iron is present in solution as a complex ferritartrate ion.

CITRATES

Heated with concentrated *sulphuric acid*, carbon monoxide is evolved and the liquid turns yellow. Carbon dioxide is also evolved as the heating is continued, and after some time the liquid blackens and some sulphur dioxide is evolved. The first action of the sulphuric acid is to withdraw the elements of water from the citric acid with formation of acetonedicarboxylic acid and carbon monoxide. The acetonedicarboxylic acid then decomposes into acetone and carbon dioxide:

$$\begin{array}{ll} CH_2COOH & CH_2COOH \\ C(OH)COOH & = CO \quad\quad +H_2O+CO, \\ CH_2COOH & CH_2COOH \end{array}$$

followed by

$$\begin{array}{ll} CH_2COOH & CH_3 \\ CO & = CO \quad +2CO_2. \\ CH_2COOH & CH_3 \end{array}$$

Citric acid differs from tartaric acid in some respects. It is less readily oxidized, and consequently does not give a silver

mirror on warming with *ammoniacal silver nitrate,* nor does it
show any positive result when treated with *hydrogen peroxide*
and a trace of a *ferrous salt.*

Calcium chloride produces no precipitate from cold neutral
solutions of citrates, but on boiling, calcium citrate separates
as a white precipitate, since the compound is less soluble in hot
water than in cold.

Citrates resemble tartrates in forming fairly stable complex
anions with certain metals such as copper and iron. These
complex cupricitrates and ferricitrates resist precipitation of the
metallic hydroxide when a source of hydroxyl ions is added.
Indeed, the ferricitrate complex is somewhat more stable than
the ferritartrate complex ion.

When citric acid is heated, it first loses its water of crystallization, then fuses, and evolves pungent fumes of aconitic acid and
other products. The alkaline and alkaline earth citrates leave a
residue of carbonate on ignition.

HYDROGEN PEROXIDE

This compound is commonly supplied in aqueous solution of
various concentrations, known as 10-volume or 20-volume
peroxide, according to the volume of oxygen which is evolved
when a definite volume of the solution is decomposed with the
aid of a catalyst such as colloidal platinum. It has both reducing
and oxidizing properties, but it is to be noted that when hydrogen
peroxide functions as a reducing agent oxygen is invariably
evolved.

Hydrogen peroxide is best recognized in solution by its reaction with a solution of *titanium dioxide* in *dilute sulphuric acid,*
which results in the formation of the orange-yellow colour of
peroxo-disulphato-titanic acid (p. 57). This reaction is preferable to the perchromic acid test (see chromium, p. 53).

Hydrogen peroxide derivatives of various acids are well
known. Examples of compounds of this kind are to be found in
perborates,* percarbonates, and also in permolybdates and

* The prefix "per" does not necessarily imply that the compound is a
derivative of hydrogen peroxide.

similar compounds. A most interesting derivative of peracetic acid was prepared long ago by Brodie, namely diacetyl peroxide, $\begin{smallmatrix}CH_3COO\\CH_3COO\end{smallmatrix}\Big|$, which was obtained by the action of barium peroxide on acetic anhydride. Barium acetate is formed at the same time:

$$BaO_2 + 2\ \begin{array}{c}CH_3CO\\ CH_3CO\end{array}\!\!\Big\rangle O = \begin{array}{c}CH_3COO\\ CH_3COO\end{array}\!\!\Big| + (CH_3COO)_2Ba.$$

The compound is dangerous to prepare on account of its explosive properties.

In examining metallic oxides which readily evolve oxygen on heating, it is important to test whether or not they are true peroxides, i.e. derivatives of hydrogen peroxide. This is best done by treating them with dilute hydrochloric acid, keeping the solution cooled in ice, filtering if necessary, and testing the resulting solution with the titanium dioxide reagent.

As an oxidizing agent, hydrogen peroxide effects the general reactions characteristic of other oxidizing agents, such as the liberation of iodine from acidified potassium iodide, the conversion of hydrogen sulphide to sulphur, and the oxidation of ferrous salts to ferric salts. In addition to this, it readily oxidizes certain sulphides to sulphates. A very useful test for the compound depending upon its capacity for oxidizing lead sulphide to lead sulphate may be performed by moistening a filter paper with a drop of a solution of lead acetate, exposing the paper to hydrogen sulphide for a moment to obtain a black spot of lead sulphide, and then adding a few drops of the solution to be tested. In the presence of hydrogen peroxide, the black spot is bleached at once:

$$PbS + 4H_2O_2 = PbSO_4 + 4H_2O.$$

Reducing reactions in which hydrogen peroxide reacts with other powerful oxidizing agents are to be found in its reactions with silver oxide, acidified potassium permanganate, and alkaline potassium ferricyanide:

$$2K_3Fe(CN)_6 + 2KOH + H_2O_2 = 2K_4Fe(CN)_6 + 2H_2O + O_2.$$

The test for reducing agents with a mixture of dilute potassium ferricyanide and ferric sulphate described on p. 107 is given by hydrogen peroxide. No alkali is added to the mixture, but reduction of the ferricyanide takes place nevertheless, the solution becoming acid with simultaneous formation of ferric ferrocyanide, resulting in the production of a fine blue colour or precipitate.

Another sensitive test, depending upon the reduction of a ferricyanide to a ferrocyanide, consists in precipitating the yellowish green cupric ferricyanide by mixing a dilute solution of copper sulphate with one of freshly dissolved potassium ferricyanide. In the presence of sodium acetate the yellowish green precipitate is readily reduced to the brown cupric ferrocyanide by traces of hydrogen peroxide.

Chapter V

SYSTEMATIC ANALYSIS OF THE METALS

PRELIMINARY EXAMINATION OF SUBSTANCES BY DRY METHODS

THE so-called dry tests for metals have been long in use, and information obtained by employing them is frequently of much value in the subsequent systematic analysis. These dry reactions were developed to a high degree of refinement by Bunsen, but are seldom applied in such an elaborate manner at the present time. Some of these tests are highly sensitive, e.g. the production of Rinmann's green as a test for zinc, and may justly be classified as microchemical reactions.

The following dry reactions should always be carried out when dealing with unknown substances.

1. Heat a small quantity of the substance in a hard glass tube. Volatile products frequently furnish information, e.g. the evolution of white fumes from ammonium salts, of oxygen from salts of oxyacids such as chlorates, bromates, and iodates, and of carbon dioxide from carbonates. Hypophosphites are decomposed with evolution of phosphine, which is highly inflammable.

2. Heat a small quantity of the substance, previously well mixed with sodium carbonate, in a cavity in a charcoal block, using the reducing flame. Compounds of easily reducible metals such as silver, lead, bismuth and antimony yield metallic beads. Copper produces spangles of the metal, but the melting point is too high for the formation of a bead. It is usually desirable to remove the bead and dissolve it in acid to identify it. Metals such as zinc, cadmium, mercury, and arsenic are immediately volatilized on reduction. In the case of arsenic, there is no difficulty about recognition, as the vapour has a highly

disagreeable odour, similar to that of garlic. Some chemists attach importance to the colour of the incrustation round a bead, but this is unnecessary in rapid work.

3. Dissolve a small quantity of the substance in water or dilute hydrochloric or nitric acid and add a little cobalt nitrate, or better, potassium cobalticyanide. Soak a filter paper in the mixed solution, and burn the paper. Examine the colour of the ash carefully. A blue ash indicates the presence of aluminium (Thénard's blue), a green ash is characteristic of zinc (Rinmann's green). Magnesium compounds produce a faintly pink ash, but the reaction is not very satisfactory. The production of these coloured products is probably due to the formation of solid solutions of cobaltous oxide with the other metallic oxide.

4. Fuse a very minute quantity of the substance in a borax bead attached to a platinum wire. The bead should be examined both in the reducing and in the oxidizing flames. Coloured glasses of sodium metaborate are produced with cobalt (deep blue), copper (pale blue), chromium (green), manganese (violet), iron (brownish) in the oxidizing flame. In the reducing flame, the copper bead ultimately assumes a ruby-red colour, due to cuprous oxide, the manganese bead becomes colourless, and the iron bead assumes a green colour, the others remain unchanged.

5. Examine the flame coloration, using a *clean* platinum wire on the substance previously moistened with dilute hydrochloric acid. Characteristic colours are produced with compounds of thallium (green), copper (vivid green), barium (apple green and very persistent), strontium (crimson), calcium (dull red), lithium (carmine), sodium (*intense* yellow), potassium (violet). The potassium flame is masked by the presence of sodium when viewed with the naked eye, but if viewed through cobalt-blue glass it appears reddish violet. It is not difficult to detect barium in the presence of other metals which produce flame colorations, as the apple green barium flame persists longer than the others.

SEPARATION OF THE METALS OF THE FIRST OR
HYDROCHLORIC ACID GROUP

Add dilute hydrochloric acid to the solution in slight but decided excess. Filter, wash the precipitate with cold water, and examine it as follows:

1. Test for *lead*. Extract the precipitate with hot water, and observe if a white crystalline precipitate separates on cooling. In any case, test the solution with potassium chromate (yellow precipitate of lead chromate) or with dilute sulphuric acid (white precipitate of lead sulphate).

2. Test for *silver* and *mercury* together after extracting the lead chloride with hot water. Pour warm dilute ammonia solution over the precipitate. Blackening indicates *mercury*. Add dilute nitric acid to the ammonia solution in slight excess. *Silver* if present is reprecipitated as the chloride. The presence of *mercury* in the blackened precipitate may be confirmed by dissolving it in dilute aqua regia and adding stannous chloride. A white precipitate turning grey (a mixture of mercurous chloride and mercury) is conclusive.

3. In ordinary work, it is unnecessary to test for other elements in the first group, but a few words may be added regarding the detection of *thallium* and *tungsten*.

Thallous chloride is very similar to *lead chloride*, but the two elements may be distinguished by the solubility of thallous sulphate in water, and by the difference in behaviour of the iodides towards sodium thiosulphate. Lead iodide is soluble in a solution of sodium thiosulphate, whereas thallous iodide is insoluble in this reagent.

Tungsten may be precipitated as yellow or white tungstic acid, readily soluble in alkalis. On adding zinc and dilute hydrochloric acid to tungstic acid, an intensely blue reduction product is obtained. Tungstic acid will not be precipitated if phosphates are present, on account of the formation of complex phosphotungstic acids.

SEPARATION OF THE METALS OF THE SECOND OR HYDROGEN SULPHIDE GROUP

Boil the diluted filtrate from the first group, and pass a *slow* stream of hydrogen sulphide through the solution for several minutes. Remove the precipitated sulphides by filtering, and ascertain at once if precipitation is complete by passing the gas once more through the filtrate. If a further precipitate is obtained, it must be added to the main precipitate.

Wash the precipitate with hot water. Then introduce it into a test tube, and digest it for about five minutes with yellow ammonium sulphide. Filter, wash, and examine the residue and filtrate separately.

First division. Sulphides insoluble in yellow ammonium sulphide. Extract the precipitate with hot dilute nitric acid and filter. Mercuric sulphide will remain undissolved, the other sulphides are converted into nitrates with separation of sulphur:

1. Test for *mercury* in the insoluble residue by dissolving it in hot dilute aqua regia and adding stannous chloride. A grey precipitate (a mixture of mercurous chloride and mercury) confirms this element.

2. Test for *lead* in the nitric acid solution by adding a little dilute sulphuric acid. Lead if present is precipitated as white lead sulphate, but the amount is likely to be small, as most of the lead will have been separated as lead chloride in the first group.

3. If lead sulphate has separated, remove it by filtering. Then add ammonia to the filtrate in slight excess to test for *bismuth*. If a white precipitate (bismuth hydroxide) separates, remove it and dissolve it in a very small quantity of hydrochloric acid. Add a considerable quantity of water. A white precipitate (bismuth oxychloride) is a satisfactory confirmation.

4. Test for *copper* and *cadmium* together in the ammoniacal filtrate from the bismuth test. The solution will be coloured blue if *copper* is present (tetrammine cation). Add potassium cyanide until the blue colour is discharged (formation of the very stable

complex anion), and pass a few bubbles of hydrogen sulphide through the colourless solution. A yellow precipitate of *cadmium sulphide* is conclusive.

Second division. Sulphides soluble in yellow ammonium sulphide. Add dilute hydrochloric acid in excess to the solution. The thio-salts of *arsenic, antimony,* and *tin* are decomposed by hydrochloric acid with reprecipitation of the sulphides of these elements. If a *white* precipitate is obtained which does not settle, it probably consists of amorphous sulphur and may be neglected. Apply confirmatory tests as follows:

1. A very small portion of the precipitate is warmed with dilute sodium hydroxide solution and aluminium turnings. Hold a piece of filter paper moistened with a drop of silver nitrate over the mouth of the tube. A black stain indicates the presence of *arsenic.* See Fleitmann's test, p. 43.

2. Boil a larger portion of the precipitate with concentrated hydrochloric acid. Reject the insoluble portion, and dilute the solution with water. Test for *antimony* and *tin* together by adding a piece of iron wire. After a short time, *antimony* if present is deposited on the wire as a brown powder. *Tin* is reduced by the iron wire from the stannic to the stannous condition. On adding mercuric chloride, the production of a grey precipitate is a satisfactory confirmation.

NOTES ON REACTIONS IN THE HYDROGEN SULPHIDE GROUP

As lead chloride is appreciably soluble in water at the ordinary temperature, some lead sulphide is likely to be precipitated in this group. Cadmium is liable to be precipitated as sulphide incompletely, on account of the readiness of the reversal of the reaction between hydrogen sulphide and hydrochloric acid. It is desirable, therefore, to test a portion of the filtrate from the group to ascertain if cadmium has been completely removed or not. In doing this, it is desirable to neutralize the filtrate partially with ammonia before applying hydrogen sulphide.

Hydrogen sulphide is a powerful reducing agent. Hence if

oxidizing substances are present in solution, these substances will be reduced with separation of sulphur, and frequently with changes of colour which correspond to changes in the state of oxidation. Thus permanganates (purple) are reduced to manganous salts (colourless), dichromates (orange) are converted into chromic salts (green or violet), ceric salts (orange) into cerous salts (colourless). Molybdates (colourless in dilute solution) are first reduced to a lower state of oxidation having a deep blue colour, but ultimately most of the molybdenum is precipitated as the brown trisulphide, MoS_3. This compound is soluble in yellow ammonium sulphide as a thiomolybdate, and hence is separated with arsenic antimony and tin. (For confirmatory tests for molybdenum see p. 48.) Vanadates (yellow) are reduced to blue vanadyl salts. Arsenates are very slowly reduced to arsenites by hydrogen sulphide, arsenious sulphide being ultimately precipitated. If the presence of an arsenate is suspected, it is desirable to effect reduction by sulphur dioxide before using hydrogen sulphide as the group reagent.

SEPARATION OF THE METALS OF THE THIRD OR AMMONIA GROUP

Boil the filtrate from the second group until the odour of hydrogen sulphide has been completely removed. Then add a few drops of nitric acid and continue the boiling for a few minutes to reoxidize certain substances, especially iron compounds, which may have been reduced by hydrogen sulphide. Then add ammonium chloride followed by a slight excess of ammonia. Filter and wash the precipitate.

1. Test a small portion of the precipitate for *phosphate* by dissolving it in dilute nitric acid and warming with a considerable excess of ammonium molybdate. If a yellow precipitate of ammonium phosphomolybdate is produced, remove phosphoric acid by the basic ferric acetate method as described below. If phosphates are absent proceed as follows:

2. Separate *iron* from *aluminium* and *chromium* by boiling the precipitate with water and a little sodium peroxide or with

sodium hydroxide and hydrogen peroxide. Ferric hydroxide remains unchanged, whereas aluminium and chromium hydroxides are dissolved as sodium aluminate and sodium chromate respectively. Filter, and examine the residue and solution separately.

3. Test the residue for *iron* by dissolving a portion of it in dilute hydrochloric acid and adding ammonium thiocyanate. An intense red colour, stable to acids but bleached by sodium fluoride or mercuric chloride, is conclusive.

4. Test the alkaline solution for *aluminium* by boiling a portion of it with ammonium chloride solution. A white gelatinous precipitate indicates *aluminium*.

5. Test a second portion of the alkaline solution for *chromium*, which will be yellow if chromate is present, with lead acetate and acetic acid. Yellow lead chromate will be precipitated.

6. In ordinary work, it is unusual to test specially for other elements, which may be precipitated as hydroxides by ammonia (beryllium, cerium, thorium, and zirconium), or for uranium (precipitated as ammonium diuranate). *Titanium* is, however, a comparatively common element, and may be recognized by dissolving a portion of the ammonia precipitate in dilute sulphuric acid and adding hydrogen peroxide. An intense orange colour (peroxo-disulphato-titanic acid) indicates the presence of this element. Methods for the separation of these elements from the commoner metals and from each other must be sought in larger works.* Much progress has, however, been made in the way of recognizing some of these metals by "spot tests". Thus beryllium may be detected by the cornflower-blue coloration which is produced with quinalizarin, and sensitive tests for the other elements are also described.

Procedure in the presence of phosphates. Dissolve the ammonia precipitate in a *small* quantity of dilute hydrochloric acid, and add excess of a moderately concentrated solution of sodium or

* E.g. *Applied Inorganic Analysis* by W. F. Hillebrand and G. E. F. Lundell, 1929, and *A System of Qualitative Analysis for the Rare Elements* by Arthur A. Noyes and William C. Bray, 1927.

ammonium acetate. Then add ferric chloride to the solution as long as a precipitate (ferric phosphate) is produced, and continue adding the reagent until the upper liquid assumes a red colour. Boil and filter. Reject the precipitate (ferric phosphate and basic ferric acetate) and test the filtrate for metals, which are precipitated in the fourth (ammonium sulphide) and fifth (ammonium carbonate) groups by the usual methods.

When phosphates are present, tests for aluminium, chromium, and iron must be carried out on portions of the original substance.

Oxalates of metals of the alkaline earths may be precipitated in the third group. They may be decomposed by ignition at a low red heat, and the resulting carbonates dissolved in dilute hydrochloric acid, and tested for barium, strontium, calcium, and magnesium by the usual methods.

Some acidic ions form stable complex ions with the metals of the third group. Thus *tartrates* form ferritartrate ions from which ammonia will not precipitate ferric hydroxide. The iron is therefore carried into the fourth group. On this account it is always desirable in a systematic analysis to ascertain what acidic constituents are present before dealing with the third group, and to vary the usual procedure as may be necessary.

Vanadium when present alone is not normally precipitated by any group reagent. It may, however, appear together with the hydroxides of the third group. In other cases, it will find its way into the ammonium sulphide group, from which it may be separated as the sulphide by acidifying the solution with hydrochloric acid.

NOTES ON SEPARATIONS BETWEEN THE AMMONIA AND THE AMMONIUM SULPHIDE GROUPS

The separation of the hydroxides of aluminium, chromium, and iron, and also of certain other metals, by ammonia in the presence of ammonium chloride from cobalt, nickel, manganese, and zinc is by no means complete. In particular, manganese is always precipitated in part together with ferric hydroxide, and cobalt is to some extent co-precipitated with aluminium hydroxide. These incomplete separations are chiefly of the nature of adsorption phenomena. In the case of manganese, co-precipitation with ferric hydroxide is facilitated because of the readiness of manganous salts to undergo oxidation in an ammoniacal solution. When the ratio of the fourth group metal to the third group metal is small, the quantity of it carried down with the ammonia precipitate may be such as to render its detection in the ammonium sulphide group a matter of difficulty. This is frequently the case with zinc. Some analysts, notably Noyes, Bray, and Spear, have abandoned the usual practice of separating the third and fourth groups, and make use of ammonium sulphide for dealing with both sets of metals together. The detailed group separations are, however, necessarily more elaborate.

A method for separating iron from manganese which is widely used when a quantitative separation is required depends upon the precipitation of the iron as basic ferric acetate. Ammonia is added cautiously to the acid solution to neutralize excess of acid, care being taken that ferric hydroxide is not precipitated. Excess of a solution of sodium or ammonium acetate is then added, and the solution boiled. All the iron separates as the insoluble basic acetate, while manganese remains in solution.

SEPARATION OF CERTAIN HYDROXIDES BY BARIUM CARBONATE

The tervalent hydroxides of aluminium, iron, and chromium, as well as the quadrivalent hydroxide of titanium, are much more weakly basic than the bivalent hydroxides of ferrous iron, cobalt, nickel, manganese, and zinc. Separation of the former from the latter is therefore effected by supplying a concentration of hydroxyl ions sufficient to precipitate the former but not the latter. A method for doing this, which has been long in use, consists in adding a suspension of *barium carbonate*, preferably freshly precipitated, to a solution of the chlorides and digesting the liquid for some time at a gentle heat. The feebly basic hydroxides are completely precipitated, carbon dioxide being evolved:

$$3BaCO_3 + 2FeCl_3 + 3H_2O = 2Fe(OH)_3 + BaCl_2 + 3CO_2.$$

The precipitated hydroxides will, of course, contain excess of barium carbonate. They may be freed from this by dissolving them in *hydrochloric acid* and reprecipitating them by *ammonia*. The more basic metals remain in solution, and may be precipitated by appropriate reagents such as *ammonium sulphide*. If it is necessary to remove the barium present in solution, a slight excess of a solution of *ammonium* or *sodium sulphate* may be added to the filtrate, and the resulting barium sulphate removed by filtering. This method of separating the hydroxides of the iron group, although slow, is efficient. It may be added that *zinc oxide* can be used instead of barium carbonate for many separations of this kind.

Separation of the Metals of the Fourth or Ammonium Sulphide Group

Add ammonium sulphide (preferably the colourless reagent*) to the filtrate from the third group, and collect and wash the precipitated sulphides. The sulphides of cobalt and nickel are black, that of manganese is pink, and zinc sulphide is white. If the precipitate is not black, there is no necessity to test for cobalt or nickel. Extract the precipitate with cold dilute hydrochloric acid. Filter, and examine the filtrate and the residue separately.

1. Test the residue for *cobalt* by a borax bead. A deep blue bead is a satisfactory indication for this element, even in the presence of much nickel.

2. Dissolve the black residue in aqua regia, and boil thoroughly in a dish until every trace of chlorine is expelled. Dilute the resulting solution with water and divide it into two parts. Apply a confirmatory test for *cobalt*, such as the α-nitroso-β-naphthol reaction or the reaction with ammonium mercuric thiocyanate, to one portion. Test the other portion for *nickel* by adding a

---

* Or pass hydrogen sulphide through the ammoniacal solution.

dilute alcoholic solution of dimethylglyoxime and ammonia. A scarlet precipitate is conclusive.

3. Boil the hydrochloric acid filtrate to expel hydrogen sulphide. Cool and add cold sodium hydroxide in excess. A precipitate insoluble in excess of the reagent is probably manganous hydroxide. Confirm for *manganese* by a fusion test with sodium carbonate and potassium nitrate (green melt of sodium manganate). Test for *zinc* in the alkaline filtrate (which will contain the metal in solution as sodium zincate) with ammonium sulphide. A white precipitate of zinc sulphide is a satisfactory indication. Alternatively the solution may be acidified and tested for *zinc* with ammonium mercurithiocyanate, which produces a white precipitate.

SEPARATION OF THE METALS OF THE FIFTH OR AMMONIUM CARBONATE GROUP

Add ammonium carbonate in excess to the filtrate from the fourth group, and warm, but do not boil, the solution. Filter and wash the precipitated carbonates and examine the precipitate as follows:

1. Dissolve the precipitate in hot dilute acetic acid. Test for *barium* by adding a slight excess of potassium chromate, which precipitates the yellow chromate. Filter off this precipitate and confirm for barium by the flame reaction.

2. Test for filtrate containing potassium chromate for *strontium* by adding dilute sulphuric acid or ammonium sulphate. A white precipitate indicates the presence of strontium. Remove the precipitate by filtering.

3. Test the filtrate after removing strontium sulphate for *calcium* by adding excess of ammonia and ammonium oxalate. A white precipitate of calcium oxalate indicates the presence of that element. Apply the flame test to confirm for calcium.

Note. On account of the similarity in properties between the three elements, there is really no satisfactory method for separating barium, strontium, and calcium by the single use of given reagents. The simple procedure described above is usually sufficient for routine work. There is no difficulty in recognizing barium in presence of either or both of the other two elements by the flame reaction, as the apple green barium coloration persists if the platinum wire is held in the flame for some time after the crimson strontium flame and the red calcium flame have vanished.

SEPARATION OF MAGNESIUM AND THE ALKALI METALS (SIXTH GROUP)

Evaporate the filtrate from the fifth group to dryness and ignite strongly to volatilize ammonium salts. Ignition must be continued until fumes cease to come off. Dissolve the residue in water and apply tests for magnesium, potassium, and sodium to separate portions of the solution.

1. Test for *magnesium* by adding ammonium chloride, ammonia, and sodium phosphate. A white crystalline precipitate (magnesium ammonio-phosphate) indicates the presence of this element.

2. Test for *potassium* either with tartaric acid or with picric acid. The former reagent precipitates potassium bitartrate, the latter yields potassium picrate. In both cases, the production of the crystalline precipitate is greatly facilitated by scratching the vessel with a glass rod.

3. Test for *sodium* with a solution of acid potassium pyroantimoniate or with dihydroxytartaric acid. Both reagents produce sparingly soluble sodium salts.

Notes. (i) Since the separation of the metals of the fifth group by ammonium carbonate is liable to be incomplete, it is highly desirable to test a portion of the soluble residue obtained after removing ammonium salts for these metals to prevent the possibility of confusion with magnesium. This may best be done by adding ammonium sulphate, ammonia, and ammonium oxalate, and filtering if necessary before testing for magnesium.

(ii) The presence or absence of *ammonium salts* should always be tested for in the original substance by boiling a portion of it with sodium hydroxide and testing for ammonia.

(iii) In exceptional cases, it may be necessary to test for *lithium*. The presence of this element is indicated in the first place by the fine carmine flame coloration. For other reactions, see p. 77. The original substance should always be examined for flame colorations.*

CONCLUDING REMARKS ON THE SEPARATION OF THE METALS

The methods which have been described of separating the metals into six groups by the successive application of group reagents are those which are commonly employed. Within the individual groups, however, various other methods are widely used, and, in particular cases, it may be very desirable to depart from the procedures which have been given. In view of the rapidly increasing number of new reagents which are being introduced into analytical work, it would seem probable that the future progress of qualitative analysis is likely to lie in the direction of simplifying separations within the groups, and applying these reagents in such a manner as to provide specific tests for particular metals.

* When a substance is to be tested for potassium in the presence of sodium by the flame reaction, it is better to use an instrument known as a "potassio-scope" rather than cobalt-blue glass. This is a small vessel having parallel glass sides and containing a dilute solution of a blue dye, and it enables the violet-red potassium flame to be easily seen by transmitted light.

Chapter VI

EXAMINATION FOR ACID RADICALS

Preliminary Experiments

UNLIKE the metallic ions, which are readily separated into some five or six groups according to their behaviour with group reagents, the acidic ions are not capable of systematic separation in this way. Attempts have been made to do this, but experience has shown that it is better, after making certain preliminary experiments, to apply specific tests for the individual acid radicals, and to confirm these by subsidiary tests. The following preliminary experiments should be made on fresh portions of the substance in all cases:

1. Add *dilute hydrochloric acid*, heating gently if necessary, and observe if a gas is evolved. If a positive result is obtained, the gas must be identified.

(*a*) *Carbon dioxide* from carbonates or cyanates. This gas is identified by its action on lime water, or better, on baryta water.

(*b*) *Sulphur dioxide* from sulphites, or with simultaneous precipitation of sulphur from thiosulphates or from hydrosulphites. Sulphur dioxide is recognized by its pungent odour and its reducing properties.

(*c*) *Hydrogen sulphide* from sulphides. It should be noted that metallic sulphides are decomposed with very varying ease by hydrochloric acid. Ferrous sulphide is at once decomposed by the dilute acid at the ordinary temperature, while antimony sulphide requires heating with the concentrated acid. Hydrogen sulphide is recognized by its odour and by its blackening effect on filter paper moistened with a solution of lead acetate.

(*d*) *Oxides of nitrogen* from nitrites. The brown fumes are easily observed and the pungent odour is characteristic. The production of a dark brown colour with a solution of ferrous sulphate is also useful.

(e) *Chlorine* from hypochlorites. The odour and bleaching action are distinctive.

Chlorine may be evolved in consequence of the oxidizing action of certain powerful oxidizing agents upon the hydrochloric acid. Persulphates, for example, when warmed with dilute hydrochloric acid liberate a mixture of oxygen and chlorine.

Some mixtures, when acidified with dilute hydrochloric acid, are decomposed with evolution of volatile products which indicate the nature of the constituents. Thus iodine is at once liberated when hydrochloric acid is added to a mixture of an iodate and an iodide, since iodic and hydriodic acids mutually decompose each other with formation of iodine and water. A mixture of an iodide and a nitrite behaves somewhat similarly, since nitrous acid oxidizes hydriodic acid to iodine. A bromide mixed with a bromate when acidified results in the liberation of bromine.

(f) *Hydrogen cyanide from cyanides.* Hydrogen cyanide or hydrocyanic acid may be recognized by its odour, which has some resemblance to that of bitter almonds. Many persons have difficulty in recognizing the odour of this gas, and in any case care must be taken to avoid inhaling it on account of its poisonous properties. If the presence of the gas is suspected, its presence should be confirmed by absorption in yellow ammonium sulphide and applying the ferric chloride test for a thiocyanate as described on p. 102. Mercuric cyanide is not decomposed by dilute hydrochloric acid unless potassium iodide is added (see p. 103).

If any positive results are obtained as indicated under any of the paragraphs (a) to (f), confirmatory tests should be applied immediately at this stage.

2. Warm a small quantity of the substance with *concentrated sulphuric acid*. Care must be taken to avoid heating the substance sufficiently strongly to cause volatilization of the sulphuric acid. Observe if a gas is evolved, and if so proceed to identify it.

(*a*) *Hydrogen chloride* from chlorides. This gas is easily recognized by its pungent odour and acidic properties.

(*b*) *Hydrogen bromide* and *bromine* from bromides. The vapour of bromine has a characteristic brownish red colour and an intensely pungent odour.

(*c*) *Iodine* with some hydrogen iodide from iodides. The violet vapour of iodine is easily observed.

(*d*) *Hydrogen fluoride* from fluorides. The gas has a pungent odour similar to that of hydrochloric acid, but if this gas is suspected, it should be tested for specially by its etching action on glass. Alternatively, a wet glass rod may be held in the tube. In the presence of hydrofluoric acid, the drop of water at the end of the tube will become coated with a white deposit of silica.

It should be noted that the chlorides of silver and mercury are not decomposed by heating with concentrated sulphuric acid.

(*e*) *Carbon monoxide* from formates, oxalates, tartrates, citrates, ferrocyanides, and ferricyanides. This gas is easily recognized by its property of burning with a blue flame. With formates and with ferrocyanides, carbon monoxide is the only gas which is evolved. With oxalates, an equal volume of carbon dioxide is also evolved. A small amount of carbon dioxide is also produced from ferricyanides. With tartrates, rapid charring takes place with simultaneous evolution of carbon dioxide and sulphur dioxide. Citrates behave similarly to tartrates, but charring takes place much more slowly.

(*f*) *Chlorine dioxide* from chlorates. This gas is extremely explosive, the action of sulphuric acid on the salt being attended with loud detonation.

(*g*) *Volatile acids*, such as nitric and acetic acids, which are stable to concentrated sulphuric acid, may be evolved. The presence of nitric acid may be recognized by adding a few copper turnings, when red fumes of oxides of nitrogen are evolved. A better method of confirming the presence of this acid is, however, to apply the "brown ring" test (see p. 88). Acetic acid may be recognized by warming the substance with concentrated sulphuric acid and a few drops of ethyl alcohol, when the fragrant odour of ethyl acetate is at once apparent.

(h) *Sulphur dioxide*. This gas may be evolved by the action of many substances which exert a reducing action upon the heated sulphuric acid. Thus with bromides, the hydrogen bromide which is first produced reduces some of the sulphuric acid to sulphur dioxide with simultaneous oxidation of the hydrogen bromide to bromine. With iodides, reduction usually goes further, with formation of sulphur and even of hydrogen sulphide. The formation of sulphur dioxide when tartrates or citrates are heated with concentrated sulphuric acid is due to the reducing action of the carbon which is produced by the charring action of the acid.

Confirmatory tests must be applied in the event of any positive results being obtained as indicated under paragraphs (a) to (h) inclusive.

It will be obvious that in the event of negative results being obtained as a result of examining the behaviour of the substance with dilute hydrochloric acid and with concentrated sulphuric acid, a large number of acid radicals have been ruled out. There are, however, certain well-known acids which show no response to either dilute hydrochloric acid or to concentrated sulphuric acid, which should be tested for at this stage. The tests may be applied directly to fresh portions of the original substance, or a solution may be prepared, free from heavy metals, by boiling the substance with sodium carbonate, filtering from the insoluble carbonates, and then testing the filtrate, which contains the acids as the sodium salts. Opinions differ as to the desirability or otherwise of making a general practice of removing heavy metals before proceeding to test for the acid radicals. In some cases the presence of heavy metals interferes with the reactions, and in such cases separation is obviously necessary, but, in the writer's opinion, the practice of *always* removing heavy metals is neither necessary nor even desirable.

3. *Test for a sulphate.* Acidify the substance in solution with

dilute hydrochloric acid, and add *barium chloride.* A white precipitate insoluble in hydrochloric acid is usually conclusive, but
it is desirable to confirm by adding dilute nitric acid and lead
nitrate or lead acetate to another portion, when a white precipitate of lead sulphate is produced.

4. *Test for a phosphate* by acidifying the substance with *nitric
acid* and warming with a large excess of *ammonium molybdate,*
when the yellow crystalline precipitate of ammonium phosphomolybdate is produced. In a systematic analysis, this test
should be carried out before proceeding to examine the metals
precipitated by ammonia, as in the event of phosphates being
present, the procedure requires modification (see p. 123). It
should be remembered that arsenates respond to ammonium
molybdate in a similar manner, but the precipitate usually only
appears on boiling, and, further, indications of the presence
of arsenic should have been obtained previously in the preliminary examinations of the substance for metals.

5. *Investigate the action of silver nitrate under varied conditions.*

(a) *In presence of nitric acid.* Insoluble precipitates are produced with chlorides (white), bromides (pale yellow), iodides
(yellow), thiocyanates (white), ferrocyanides (white), ferricyanides (orange), nitroprussides (pink).

(b) *In presence of acetic acid.* The silver salts of certain acids
are readily soluble in dilute nitric acid, but insoluble in acetic
acid. Oxalates and iodates produce white precipitates under
these conditions. Chromates or dichromates precipitate red
silver chromate.

(c) *In neutral solution.* A considerable number of acidic ions
yield precipitates which are not obtained in acid solution.
Phosphates produce yellow silver phosphate, and arsenates the
chocolate-coloured silver arsenate. White precipitates indicate
the presence of borates, oxalates, tartrates, citrates, and other
organic acids. Specific tests must be applied for confirmation in
all cases.

Preparation of a neutral solution. Remove heavy metals as
insoluble carbonates, by boiling the substance with excess of

sodium carbonate and filtering. Add dilute nitric acid to the filtrate, drop by drop, and boil gently to expel carbon dioxide until the solution is acid to litmus paper. Then add very dilute ammonia solution until a solution strictly neutral to both red and blue litmus paper is obtained. This must be done with care, as the slightest excess of either acid or alkali will involve failure of the tests. The solution should be divided into three parts, one part being tested with silver nitrate, and the other two parts reserved for testing with calcium chloride and ferric chloride, as explained under sections 6 and 7 below.

6. *Calcium chloride* is added to the *neutral solution*. Tartrates produce a white precipitate in the cold, citrates are precipitated on boiling. A number of other acids also yield sparingly soluble calcium salts, and confirmatory tests must be applied in all cases.

7. *Ferric chloride* is added to the *neutral solution*. A dark red colour indicates the presence of an acetate, formate, or thiocyanate. Ferric acetate and ferric formate are hydrolysed on boiling the solutions with separation of insoluble basic salts. Formates may be distinguished from acetates by applying tests for reducing agents, positive results being obtained with the former and negative results with the latter. Ferric thiocyanate is much darker in colour than ferric acetate or formate, and its colour is not destroyed by acids, but is bleached by adding either mercuric chloride or sodium fluoride. If a yellowish precipitate is obtained, the presence of a phosphate is indicated.

After the substance has been examined in the manner indicated above, specific confirmatory tests must be applied to verify the presence of any acid radicals which have been indicated. This is particularly important, because the acidic ions, unlike those of the metals, are usually compounds, and therefore are capable of undergoing decomposition in the presence of other substances.

Remarks on the Detection of More than One Acid Radical

When more than one acid radical has to be detected, although no new principle is concerned, the problem may present difficulties for various reasons. Thus if the properties of the acidic ions are similar, and respond to the same reagents, special methods may be necessary (compare the section on the detection of the halides in presence of each other). Again, a reaction which is perfectly satisfactory for some particular acid radical may fail in the presence of another. Another source of difficulty may arise if two acidic ions unite to form complex ions in which the reactions of both constituents are wholly or partially obscured. The problems to be encountered may best be illustrated by reference to some simple examples.

Nitrate and nitrite. The detection of small quantities of a nitrate in the presence of a nitrite is more difficult than the converse problem. There are a number of sensitive tests for nitrites, e.g. the guaiacol reaction, which are unaffected by nitrates. But it will be obvious that the "brown ring" test for a nitrate cannot be applied in the presence of a nitrite, as the whole solution containing ferrous sulphate would become dark brown on adding sulphuric acid. The best method of testing for a nitrate in such circumstances would be to destroy the nitrous acid by adding hydrazine sulphate, and then apply the "brown ring" test.

Arsenate and phosphate. Both of these ions respond similarly to the ammonium molybdate test. In neutral solution, arsenates produce a brick-red precipitate with silver nitrate, while phosphates produce a yellow precipitate. A mixture of the two would produce a precipitate of altogether indefinite colour with silver nitrate. The best method of testing for an arsenate in the presence of a phosphate would be to reduce the arsenate to arsenite by sulphur dioxide, and then precipitate the arsenic with hydrogen sulphide, or if the quantity of arsenic is very small to apply Fleitmann's test or Reinsch's test to the reduced solution.

Carbonate and bicarbonate. Calcium carbonate is practically insoluble in water, whereas calcium bicarbonate is soluble. Add excess of a solution of calcium chloride to the liquid, filter off the calcium carbonate, then add a few drops of ammonia to the clear solution. Calcium carbonate will be precipitated as a result of the neutralization of the bicarbonate.

Detection of sulphite in presence of thiosulphate. Aldehydes react with sulphites in aqueous solution with formation of bisulphite derivatives and production of an alkaline solution:

$$R \cdot CHO + Na_2SO_3 + H_2O = R \cdot CH(OH)SO_3Na + NaOH.$$

Thiosulphates do not react with aldehydes. To the neutral solution add a little formaldehyde or acetaldehyde and a few drops of a solution of phenolphthalein. If a red colour develops, the presence of a sulphite may be inferred.

The detection of a number of the sulphur acids in a mixture may sometimes be difficult. For details, we must refer the reader to larger works.

DETECTION OF THE HALIDES IN PRESENCE OF EACH OTHER

The problem of detecting chlorides in the presence of bromides or iodides arises less frequently than that of detecting either bromides or iodides in the presence of chlorides. The method of detecting chlorides in the presence of the other halides which is almost universally employed is that of distillation of the solid substance, previously mixed with finely powdered potassium dichromate, with concentrated sulphuric acid. The volatile products are passed into water. If chlorine is present, it volatilizes as chromyl chloride, CrO_2Cl_2, which is at once hydrolysed to a mixture of hydrochloric and chromic or dichromic acids (see p. 81). Bromine and iodine volatilize unchanged.

There are several methods for detecting traces of bromides or iodides or both in the presence of chlorides which may now be considered. These various methods depend ultimately upon two main principles, viz. first, the varying degrees of insolubility of some metallic salts, such as the silver or occasionally of the

thallous salts, of which the chlorides are the least insoluble and
the iodides the most insoluble, and secondly, the differential
behaviour of the halides towards oxidizing agents, which really
depends upon the varying degree of electronegative character
of the halogens.

1. Ladenburg's method for detecting traces of iodides in the
presence of chlorides or bromides consists in adding ammonia
to the solution followed by silver nitrate. Pale yellow silver
iodide is precipitated under these conditions, since silver iodide
is practically insoluble in a solution of ammonia. The tendency
to form complex cations of the type $Ag(\overset{+}{N}H_3)_2$ is least with the
iodide.

2. Thallous nitrate added to a solution of a mixture of the
halides precipitates thallous iodide preferentially to the bromide
and the bromide preferentially to the chloride. The order of
solubility of the thallous halides is the same as that of the silver
halides, thallous chloride being some ten times as soluble as the
bromide, and thallous bromide about ten times as soluble as
thallous iodide.

3. If chlorine water is added to a dilute solution containing
an iodide and a bromide, the iodine, being the least electronegative halogen, is liberated first. On shaking the liquid with
an organic solvent such as chloroform, the latter assumes a
violet colour due to the dissolved iodine. On adding more
chlorine water the iodine is oxidized to iodine monochloride,
which is colourless. Further addition of chlorine water results
in the liberation of bromine, which on shaking dissolves in the
chloroform layer with production of a brown colour (see p. 83).
It may be remarked that iodine monochloride is readily hydrolysed to hypoiodous acid, unless sufficient hydrochloric acid is
present to suppress the hydrolysis, but this does not interfere
with the test since the hypoiodous acid would decompose into
hydriodic and iodic acids, and these acids would mutually
decompose with separation of iodine. This iodine would then be
oxidized to iodic acid by adding more chlorine water.

4. Numerous methods of using oxidizing agents differentially

towards bromides and iodides have been suggested. One
method, due to Richards, consists in adding hydrogen peroxide
and acetic acid and heating the solution to boiling, iodine being
completely removed by this treatment. On adding more
hydrogen peroxide and dilute sulphuric acid, bromine is slowly
liberated on boiling and can be recognized by allowing the
vapours to react with filter paper impregnated with a solution of
fluorescein (see p. 83). This method of using hydrogen peroxide
differentially enables a very satisfactory separation to be effected.
For most ordinary work, it may be conveniently supplemented
with confirmatory tests involving the use of silver nitrate.
A solution containing the three halides, heavy metals being
absent, is examined as follows:

The dilute solution is first tested for iodide by adding ammonia and silver nitrate to a test portion, when the yellow
precipitate of silver iodide is at once apparent.

A larger test portion is boiled with hydrogen peroxide and
acetic acid. Iodine volatilizes in the steam, and the liquid
becomes colourless. If a portion of the liquid is now tested with
silver nitrate and ammonia, a precipitate of silver bromide,
having an extremely pale yellow colour, is observed.

Dilute sulphuric acid and a little more hydrogen peroxide is
now added to the residual liquid, which is gently boiled for
about a quarter of an hour. The decomposition of the bromide
proceeds slowly, but should be completed in this time. The
progress of the decomposition may be judged by holding a filter
paper, previously impregnated with fluorescein, in the steam.
If the liquid thus treated be tested with silver nitrate and
ammonia, no precipitate should be produced, since silver
chloride is readily soluble. On adding a slight excess of nitric
acid to the ammoniacal liquid, however, a precipitate of silver
chloride will at once appear.

THE EXAMINATION OF OXIDES

When it is concluded that a substance is an oxide, experiments should be made with the object of referring the oxide to its appropriate class. In this connexion, it must be remembered that the terms basic oxide and acidic oxide are largely relative. Many oxides are known which are of an amphoteric character. "Higher" oxides may be true peroxides, i.e. derivatives of hydrogen peroxide, or so-called polyoxides, like manganese dioxide, which evolve chlorine when warmed with hydrochloric acid, but are not peroxides in the stricter sense. The following experiments are suggested with the object of classifying the various oxides.

1. Warm the compound with hydrochloric acid. If the substance dissolves without any evolution of a gas it is probably a basic oxide. If chlorine is evolved, the substance is a "higher" oxide of the lead dioxide or manganese dioxide type, or it may be an acidic oxide of a metal.

2. Warm the substance with a solution of sodium carbonate. If the compound dissolves with evolution of carbon dioxide, it is an acidic oxide (metallic or non-metallic).

3. Grind the substance with water and pour the mixture into *cold* dilute hydrochloric acid. Test a small portion of the clear liquid with a solution of titanium dioxide in dilute sulphuric acid. If an orange or yellow colour is produced, it may be concluded that this is due to hydrogen peroxide, and therefore that the oxide is a true peroxide.

4. If the oxide is soluble in both acids and alkalis, it must obviously be classified as amphoteric.

CLASSIFICATION OF SUBSTANCES AS OXIDIZING OR REDUCING AGENTS

Although the terms oxidizing agent and reducing agent are relative rather than absolute, it is sometimes desirable to make experiments, such as are indicated below, for the purpose of classifying substances with reference to their characteristic behaviour.

(a) Oxidizing agents.

1. The liberation of iodine from acidified potassium iodide. Powerful oxidizing agents may also liberate bromine from hydrobromic acid or chlorine from hydrochloric acid.

2. The separation of sulphur from hydrogen sulphide.

3. The conversion of ferrous salts into ferric salts. This test is best carried out with the aid of ammonium thiocyanate as an indicator (see p. 50).

(b) Reducing agents.

1. The formation of silver in mirror condition by interaction of the substances with ammoniacal silver nitrate. Excess of ammonia should be avoided, and the interior of the vessel must be scrupulously clean (p. 30).

2. The production of a red precipitate of cuprous oxide by warming the substance with an alkaline cupritartrate or cupricarbonate solution (p. 38).

3. The conversion of ferric salts into ferrous salts. This test is best carried out with the aid of potassium ferricyanide as an indicator (see pp. 51 and 107).

4. The conversion of dichromates (orange) into chromic salts (green or violet), see pp. 8 and 122.

5. The conversion of permanganates (purple) into colourless manganous salts. For other colour changes which depend on reduction, see pp. 8 and 122.

Both classes of substances are capable of bleaching dyestuffs. Thus azo dyes are bleached by halogens, and hydrosulphites convert methylene blue into the leuco compound.

TABULAR SUMMARY OF THE GROUP REACTIONS OF THE METALS ✿

Group reagent	Metals precipitated	Remarks
Dilute HCl (see p. 119)	Ag, Hg(ous), Pb, *Tl* as chlorides *W* as tungstic acid	Precipitates are white, except tungstic acid which is usually yellow
H_2S in presence of dilute HCl (see pp. 120 and 121)	Pb, Hg(ic), Bi, Cu, Cd as sulphides which are insoluble in yellow ammonium sulphide. As, Sb, Sn, and *Mo* as sulphides which are soluble in yellow ammonium sulphide	CdS, As_2S_3, and SnS_2 are yellow, Sb_2S_3 is orange, SnS is brown. The other sulphides are black. Colour changes may take place in the solution in consequence of reducing action of H_2S (see p. 122)
NH_3 in presence of NH_4Cl (see pp. 122, 123 and 124)	Al, Cr, Fe(ic) as hydroxides. Also *Be, Ti, Ce, Th,* and *Zr* as hydroxides *U* as $(NH_4)_2U_2O_7$	$Al(OH)_3$ and the hydroxides of the rarer metals are white. $Cr(OH)_3$ is green. $Fe(OH)_3$ is reddish brown. For phosphates see pp. 123 and 124
$(NH_4)_2S$ in ammoniacal solution (see pp. 126 and 127)	Co, Ni, Mn, Zn as sulphides	CoS and NiS are black. MnS is pink and ZnS is white *V* when present alone may be separated from this group as the sulphide by acidifying the solution (see p. 124)
$(NH_4)_2CO_3$ in ammoniacal solution (see p. 127)	Ba, Sr, Ca as carbonates	Precipitates are white
Na_2HPO_4 in presence of NH_3 and NH_4Cl (see p. 128)	Mg as the ammonio-phosphate	Precipitate is white. The alkali metals are not precipitated by a group reagent

* For preliminary examination of substances by dry methods, see pp. 117 and 118.

TABULAR SUMMARY OF THE PRELIMINARY REACTIONS
OF THE ACID RADICALS

Reagent	Gas evolved or other effect produced	Indications and Remarks
Dilute HCl (see pp. 130 and 131)	CO_2, SO_2, H_2S, N_2O_4, Cl_2, HCN may be evolved. S may be precipitated with evolution of SO_2	Carbonates, sulphites, sulphides, nitrites, hypochlorites, cyanides, thiosulphates, hydrosulphites
Concentrated H_2SO_4 (see pp. 131, 132 and 133)	HCl, HBr with Br_2, I_2, HF, CO alone or with CO_2, ClO_2, also volatile acids, and sometimes SO_2 may be evolved also	Chlorides, bromides, iodides, fluorides, formates, oxalates, tartrates, citrates, ferrocyanides, ferricyanides, chlorates, nitrates, acetates. *Note*: Borates may be tested for by the ethyl borate test (see p. 100)
$BaCl_2$ in presence of dilute HCl (see pp. 93, 133 and 134)	White precipitate	Sulphates. Other Ba salts are soluble in acids
$(NH_4)_2MoO_4$ in presence of HNO_3 (see p. 134)	Yellow crystalline precipitate on heating	Phosphates or arsenates. Arsenate requires boiling
$AgNO_3$ in presence of dilute HNO_3 (see p. 134)	Precipitates (white, pale yellow, yellow, orange, or pink)	Chlorides, bromides, iodides, thiocyanates, ferrocyanides, ferricyanides, nitroprussides
$AgNO_3$ in presence of CH_3COOH (see p. 134)	White precipitate Dark red precipitate	Oxalates or iodates Chromates or dichromates
$AgNO_3$ in neutral solution (see p. 134)	Yellow precipitate Brick red precipitate White precipitate	Phosphates Arsenates Borates, oxalates, tartrates, citrates, etc.
$CaCl_2$ in neutral solution (see p. 135)	White precipitate in the cold *Ibid.* on boiling	Tartrates Citrates
$FeCl_3$ in neutral solution (see p. 135)	Dark red colour Yellowish precipitate	Acetates, formates, thiocyanates (stable in acid solution) Phosphate

Separation of the Metals of the First Group

To a solution of the substance, prepared as described on pp. 19 and 20, add dilute hydrochloric acid in slight excess.

FILTER

RESIDUE. The precipitate may consist of the chlorides of silver, mercury(ous) and lead.
Wash the precipitate with cold water, reject the washings. Then extract the precipitate with hot water.

FILTRATE. *This will contain the metals of the second and subsequent groups.*

FILTER

RESIDUE. This will consist of silver chloride and mercurous chloride. Digest with warm dilute ammonia.

FILTRATE. If a white crystalline precipitate separates, lead is indicated. In any case confirm by testing the solution with potassium chromate (yellow precipitate) or with dilute sulphuric acid (white precipitate).

FILTER

RESIDUE. If black, mercury is indicated. Confirm by dissolving in dilute aqua regia and adding stannous chloride. A grey precipitate verifies mercury.

FILTRATE. This will contain silver as the ammine salt. Add dilute nitric acid. A white precipitate confirms silver.

Note. For the detection of thallium and tungsten, see p. 119.

Separation of the Metals of the Second Group

Dilute the filtrate from the first group with water, heat to boiling, and pass a *slow* stream of hydrogen sulphide through the solution until precipitation is complete (see p. 120).

Filter

Residue. This may consist of the sulphides of mercury(ic) lead, bismuth, copper and cadmium (all insoluble in yellow ammonium sulphide) and also of arsenic, antimony, and tin (all soluble in yellow ammonium sulphide).

Wash the precipitate with hot water, and digest it for about 5 min. with yellow ammonium sulphide.

Filtrate. *This contains the metals of the third and later groups.*

Filter

Residue (first division, or subgroup A). Sulphides of mercury(ic), lead, bismuth, copper and cadmium.

Wash with hot water, and digest with hot dilute nitric acid.

Filtrate (second division or subgroup B). *This will contain the sulphides of arsenic, antimony and tin, as thiosalts. See p. 146 for the separation.*

Filter

Residue. This will consist of mercuric sulphide and sulphur. Dissolve in hot dilute aqua regia and add stannous chloride. A grey precipitate indicates **mercury.**

Filtrate. This will consist of lead, copper, bismuth, and cadmium as nitrates. Test for lead by adding dilute sulphuric acid, and if lead sulphate separates, filter it off. Add excess of ammonia.

Filter

Residue. Probably bismuth hydroxide. Dissolve in the least quantity of hydrochloric acid, and add much water. A white precipitate of the oxychloride indicates **bismuth.**

Filtrate. A blue solution indicates **copper.** This may be confirmed by adding excess of acetic acid and potassium ferrocyanide to a portion of the solution. A brown precipitate is conclusive. To the remainder of the alkaline solution add potassium cyanide till the liquid is colourless, and pass hydrogen sulphide through it. A yellow precipitate indicates **cadmium.**

SEPARATION OF THE METALS OF THE SECOND GROUP (SUB-GROUP B) (SULPHIDES SOLUBLE IN YELLOW AMMONIUM SULPHIDE)

Acidify the filtrate obtained by digesting the sulphides with ammonium sulphide with excess of dilute hydrochloric acid.

FILTER

| RESIDUE. Sulphides of arsenic, antimony, and tin. Test a small portion for arsenic by warming with aluminium turnings and sodium hydroxide. Hold a filter paper moistened with a few drops of silver nitrate over the mouth of the tube. A black stain indicates arsenic (see p. 43). Boil the remaining portion with concentrated hydrochloric acid for several minutes. Filter, and test the liquid for arsenic and antimony by diluting with water and leaving a piece of iron wire in the solution for a few minutes. Antimony is left on the wire as a brown powder. Test the clear liquid by adding mercuric chloride. If tin is present a grey precipitate is produced. | *Reject the filtrate* |

Note. If tin is present originally in the stannous condition it will separate as *stannic* sulphide when the ammonium sulphide solution is acidified, and will dissolve in concentrated hydrochloric acid as *stannic* chloride. Iron wire will reduce it to the stannous condition, and a grey precipitate will thus be obtained on the addition of mercuric chloride.

SEPARATION OF THE METALS OF THE THIRD* GROUP IN THE ABSENCE OF PHOSPHATES

Boil the filtrate from the second group to expel hydrogen sulphide, add a few drops of nitric acid, and continue the boiling for a short time. Then add ammonium chloride and ammonia until the solution is alkaline. Heat to boiling.

FILTER

RESIDUE. Wash the precipitate carefully, and apply the test with nitric acid and ammonium molybdate to a small portion to ascertain the presence or absence of phosphates. Boil the main portion of the precipitate, which may consist of the hydroxides of iron, aluminium, and chromium, with a little sodium peroxide or with hydrogen peroxide and sodium hydroxide.	FILTRATE. *This will contain the metals of the fourth and later groups.*

FILTER

RESIDUE. Probably ferric hydroxide. Dissolve in dilute hydrochloric acid and add ammonium thiocyanate. A very dark red colour confirms iron.	FILTRATE. This will contain aluminium and chromium as anions. Divide the solution into two parts. Boil one part with ammonium chloride—a white gelatinous precipitate confirms aluminium. Add acetic acid and lead acetate to the other part—a yellow precipitate confirms chromium.

Note. For the detection of rare elements in the ammonia precipitate, see p. 123.
* Sometimes known as Group III, A.

Separation of the Metals of the Third Group when Phosphates are Present

Dissolve the precipitate obtained by ammonia, after washing with hot water, in a *small* quantity of dilute hydrochloric acid. Then add excess of a fairly concentrated solution of ammonium or sodium acetate, followed by ferric chloride. Continue adding ferric chloride until precipitation ceases and the upper liquid becomes red. Boil for a short time.

Filter

Residue. This will consist of ferric phosphate, and possibly chromium and aluminium as phosphates, together with basic ferric acetate. Apply confirmatory tests for aluminium and chromium. Iron must be tested for in the original substance.

Filtrate. Add ammonium chloride and ammonia, and pass hydrogen sulphide through the solution. If a precipitate appears, apply tests for the metals of the fourth group. If not add ammonium carbonate and examine the precipitate thus obtained for the metals of the fifth group.

Notes. (1) If too much hydrochloric acid is added to dissolve the ammonia precipitate, ferric chloride will not precipitate the phosphates, even in presence of ammonium or sodium acetate. This may be obviated by partial neutralization of the solution with sodium carbonate before adding the acetate.

(2) Other methods for eliminating phosphates have been described. Thus Smith (*Journ. Chem. Soc.* 1933, p. 253) has described a method in which the buffering is effected with the use of sodium formate instead of acetate, and Gonward and Ward (ibid. 1937, p. 1337) have recommended the removal of phosphates as ammonium phosphomolybdate. For details, the original papers must be consulted.

(3) Oxalates of the alkaline earth metals may be precipitated in the ammonia group. These may be decomposed by heating to low redness and dissolving the resulting carbonates in hydrochloric acid and proceeding as described on p. 124.

SEPARATION OF THE METALS OF THE FOURTH* GROUP

To the filtrate from the third group add ammonium sulphide, or pass hydrogen sulphide through the ammoniacal solution. This alternative procedure has certain advantages.

FILTER

RESIDUE. This may contain cobalt, nickel, manganese, and zinc as sulphides.

Wash the precipitate carefully. Extract it with cold dilute hydrochloric acid.

FILTRATE. *This will contain the metals of the fifth and sixth groups.*

FILTER

RESIDUE. Cobalt and nickel sulphides. Test with a borax bead. A deep blue bead is conclusive for cobalt, even in presence of much nickel. Test for nickel by dissolving the precipitate in boiling dilute aqua regia. Continue boiling until all chlorine is expelled. Then add excess of ammonia and a solution of dimethylglyoxime. A scarlet precipitate confirms nickel.

FILTRATE. This will contain manganese and zinc. Boil to expel hydrogen sulphide. Cool, and add cold sodium hydroxide.

FILTER

RESIDUE. Probably manganous hydroxide. Dissolve in dilute hydrochloric acid and add excess of ammoniacal silver nitrate. A black precipitate confirms manganese.

FILTRATE may contain zinc as sodium zincate. Pass hydrogen sulphide. A white precipitate confirms zinc. Alternatively zinc may be confirmed after acidifying the solution by the mercurithiocyanate test (pp. 70 and 127).

* Sometimes known as Group III, B.

150

Separation of the Metals of the Fifth Group

To the filtrate from the fourth group add ammonium carbonate. Warm, but do not boil, the solution.

FILTER

RESIDUE. This may consist of calcium, strontium, and barium as carbonates.

Wash the precipitate, dissolve it in hot dilute acetic acid, and add potassium chromate in slight excess.

FILTER

RESIDUE. Probably barium chromate. Confirm by the flame test. A persistent apple green flame indicates **barium.**

FILTRATE. This will contain strontium and calcium. Divide the solution into two parts. To one part add calcium sulphate. A white precipitate after a time indicates strontium. Confirm by the flame reaction—crimson indicates **strontium.** To the other part add dilute sulphuric acid and filter off strontium sulphate (if present). Add ammonium chloride, ammonia, and ammonium oxalate. A white precipitate indicates calcium. Confirm by the flame test—dull red indicates **calcium.**

FILTRATE. *This will contain magnesium and other metals of the sixth group.*

Separation of the Metals of the Sixth Group

Divide the filtrate obtained from the fifth group into two parts. To one part add sodium phosphate. A white crystalline precipitate indicates **magnesium**.

Evaporate the other part to dryness and ignite strongly until ammonium salts have been completely volatilized. Examine the solid by the flame reaction. An intense yellow flame indicates **sodium**. A violet flame indicates **potassium**. If both metals are present, the violet potassium flame will be obscured, but if viewed through blue glass, the reddish violet flame of potassium will be clearly visible.

Dissolve the residue in water and apply specific tests for potassium, e.g. the picric acid test or the tartaric acid test.

For the detection of lithium, see p. 129.

Notes. (1) The precipitation of the alkaline earth metals by ammonium carbonate is liable to be incomplete, and traces of these metals are therefore liable to be carried from the fifth group into the sixth group, and consequently mistaken for magnesium. It is therefore wise to test with ammonium chloride, ammonia, and ammonium oxalate, and filter if a precipitate separates, before adding sodium phosphate in testing for magnesium.

(2) A valuable confirmatory test for magnesium is the blue adsorption product obtained with paranitrobenzeneazoresorcinol. This test must be applied in the absence of ammonia and ammonium salts (see p. 75).

(3) Ammonium salts should always be tested for in the original substance by boiling with aqueous sodium hydroxide, when the evolution of ammonia is easily recognized.

THE EXAMINATION OF SUBSTANCES WITH INDICATORS

Information regarding the nature of many substances may frequently be obtained by examining the behaviour of their aqueous solutions towards indicators. Thus it is possible to distinguish between salts, such as sodium bicarbonate and sodium carbonate, or the two hydrogen sodium phosphates, by testing solutions prepared from them with indicators which differ in sensitiveness. All that is necessary is to dissolve a small quantity of the solid in distilled water, divide the solution into two or three parts, and test each part with a drop of a solution of a different indicator, chosen from a list such as is shown below. Strict neutrality at room temperature corresponds to a pH value of 7; values numerically less than this are acid, those which are greater than 7 numerically are alkaline.

Indicator	Colour change		pH range
	Acid	Alkaline	
Methyl orange	red	yellow	3·0– 4·4
Bromphenol blue	yellow	blue	3·0– 4·6
Bromcresol green	yellow	green	3·6– 5·4
Methyl red	red	yellow	4·4– 6·2
Bromcresol purple	yellow	purple	5·1– 6·7
Bromthymol blue	yellow	blue	6·0– 7·6
Cresol red	yellow	violet red	7·1– 8·8
* Thymol blue	red	yellow	1·2– 2·8
	yellow	blue	8·0– 9·6
Phenolphthalein	colourless	red	8·2–10·0
Thymolphthalein	colourless	blue	9·3–10·5

* This indicator has a double colour range.

Index

Titanium, 56, 57
   precipitation of in ammonia group,
   123
Tungsten, 36, 37
   precipitation of in hydrochloric acid
   group, 119

Uranium, 62
   precipitation of in ammonia group,
   123

Vanadium, 71, 72
   behaviour of in systematic analysis,
   122, 124

Zinc, 69, 70
   separation of, 126, 127
Zirconium, 61
   precipitation of in ammonia group,
   123

Printed in the United States
By Bookmasters